Discriminant Analysis

DISCRIMINANT ANALYSIS

Peter A. Lachenbruch

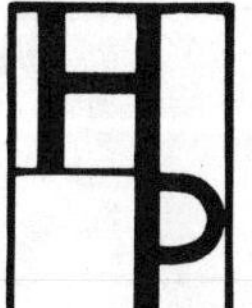

HAFNER PRESS
A Division of Macmillan Publishing Co., Inc.
New York
Collier Macmillan Publishers
London

Hafner Press
A Division of Macmillan Publishing Co., Inc.
866 Third Avenue, New York, N.Y. 10022

Collier Macmillan Canada Ltd.

Library of Congress Cataloging in Publication Data

Lachenbruch, Peter A
Discriminant analysis.

Bibliography: p.
1. Discriminant analysis. I. Title.
QA278.65.L32 519.5'3 74-11057

ISBN 0-02-848250-6

Printed in the United States of America

1 2 3 4 5 6 7 8 9 10—79 78 77 76 75

Preface

This introduction to applied discriminant analysis and problems related to the subject is written for those who use discriminant analysis in their work, for those who wish a brief summary of the state-of-the-art, and for those consumers of statisticians' services who wish a deeper knowledge of this subject. Those who want a survey of theoretical developments will not find it here; the recent paper by Das Gupta (*142a*) should serve their needs.

In recent years, a great deal of work has been done by engineers in the areas of character recognition and pattern recognition. Unfortunately, much of this work has not been noticed by statisticians, and occasional duplication of effort has resulted. A number of references to this work are included in the bibliography.

The mathematical and statistical level of this book is fairly elementary. It is assumed that the reader has a knowledge of such basics as distributions and elementary probability concepts. A knowledge of matrix algebra is required for some derivations. However, it is hoped that the discussions can be followed without this knowledge.

The book is organized as follows. Chapter 1 gives a general introduction to the subject of discriminant analysis with the basic derivations. Chapter 2 discusses evaluation of discriminant functions. Chapter 3 is concerned with robustness properties of linear discriminant functions. Some nonparametric techniques are covered in Chapter 4. The multiple-group problem is discussed in Chapter 5. Other methods, are covered in Chapter 6. These include sequential methods, variable selection, and Bayesian methods. A fairly extensive bibliography is included.

In any work of this type, the author is indebted to a large number of people who have eased the way. Many of these had their influence indirectly. I offer thanks to these anonymous ones. For first stimulating my interest in this fascinating area of statistics, I thank Jean Dunn and Ray Mickey. I owe much to Bernie Greenberg who, as Head of the Department of Biostatistics at the University of North Carolina, helped me grow as a statistician.

This book was written during a sabbatical leave at the University of California, Berkeley, School of Public Health. I wish to thank my colleagues there for the encouragement I received from them. Chin Long Chiang, in particular, went out of his way to solve administrative problems. The contents of this book were tried out on a class in the Spring quarter, 1974. The members of the class made valuable comments on the contents of the book. Larry Kupper read an early draft of the manuscript and made valuable comments on it. Bonnie Hutchings typed the manuscript with care and a tender touch.

Work on this book was supported by Research Career Development Award HD-46344. My wife, Ella, and son, Jerry, have made this year an easy, enjoyable and productive one for me. Finally, I want to thank my parents who provided a loving home and intellectual stimulation as I grew up.

Contents

Terms

Term	Definition
$\mathbf{x}$	$k \times 1$ vector observation
Π_i	population i
$\boldsymbol{\mu}_i$	$k \times 1$ mean vector in Π_i
$\boldsymbol{\Sigma}_i$	$k \times k$ covariance matrix in Π_i
$\bar{\mathbf{x}}_i$	$k \times 1$ sample mean vector in Π_i
$\mathbf{S}_i$	$k \times k$ sample covariance matrix in Π_i
p_i	a priori probability that an observation comes from Π_i
k	number of variables
$D_T(\mathbf{x}) = (\mathbf{x} - \frac{1}{2}(\boldsymbol{\mu}_1 + \boldsymbol{\mu}_2))'\boldsymbol{\Sigma}^{-1}(\boldsymbol{\mu}_1 - \boldsymbol{\mu}_2)$	true discriminant function (parameters known)
$D_S(\mathbf{x}) = (\mathbf{x} - \frac{1}{2}(\bar{\mathbf{x}}_1 + \bar{\mathbf{x}}_2))'\mathbf{S}^{-1}(\bar{\mathbf{x}}_1 - \bar{\mathbf{x}}_2)$	sample discriminant function (parameters unknown)
$f_i(\mathbf{x})$	density function of $\mathbf{x}$ in Π_i
P_i	probability of misclassifying an observation from Π_i
Φ	cumulative normal distribution function
n_i	sample size from Π_i
δ^2	$(\boldsymbol{\mu}_1 - \boldsymbol{\mu}_2)'\boldsymbol{\Sigma}^{-1}(\boldsymbol{\mu}_1 - \boldsymbol{\mu}_2) =$ Mahalanobis δ^2 distance (parameters known)
D^2	$(\bar{\mathbf{x}}_1 - \bar{\mathbf{x}}_2)'\mathbf{S}^{-1}(\bar{\mathbf{x}}_1 - \bar{\mathbf{x}}_2) =$ Mahalanobis D^2 distance (parameters unknown)
g	number of populations

1 *Basic ideas of discriminant analysis*

The basic problem of discriminant analysis is to assign an observation, **x**, of unknown origin to one of two (or more) distinct groups on the basis of the value of the observation. In some problems fairly complete information is available about the distribution of **x** in the two groups. In this case we may use this information and treat the problem as if the distributions are known. In most cases, however, the information about the distribution of **x** comes from a relatively small sample from the groups, and slightly different procedures are used. Other problems arise in practical applications of the discriminant procedures:

1. How well does the assignment rule perform?

2. How robust is the rule to departure from the assumptions that are made?

3. What variables should be selected for use in the assignment rule?

This book will be concerned with these and other problems.

Some authors view discriminant analysis as a technique for the description and testing of between-group differences. The tests involved are identical with those of multivariate analysis of variance (MANOVA). This point of view may be found in Cooley and Lohnes (*117*). It is my view that discriminant analysis is concerned with the problem of assigning an unknown observation to a group with a low error rate. The function or functions that are used to do that assignment may be identical to the ones used in the MANOVA procedures.

In the background of the discriminant analysis problem, there is the assumption that somehow we are able to classify the initial data correctly. That is, in defining the groups, some variable or variables exist that allow us to establish the groups. For example, in a study of lung cancer, a biopsy would be used to define the groups "cancer" and "noncancer." In a study to predict survival or nonsurvival following a heart attack, one simply observes the patients after a period of time and divides them into survivors and nonsurvivors. These variables cannot be easily used to predict the group to which a patient belongs. In the lung cancer example, because of the expense and discomfort involved, one would prefer to operate only on those patients who are highly likely to have the disease; in the heart attack case, the time lag makes it impossible to tell which group the patient belongs to until after the patient has died or survived. Thus one will use other variables, which are less costly and available at the time the patient appears. Hopefully, these variables will be sufficiently sensitive to allow a fairly accurate assignment to be made. In addition, if early diagnosis of a disease is made, future treatment may alter group membership, and the groups must be redefined later. (For example, early prognosis of death in the heart attack example might lead to earlier and better treatment and permit some patients to survive who otherwise would have died.)

We now give several examples of types of discriminant analysis problems.

1. The trace of an electrocardiogram is divided into 5-millisecond intervals, and a reading is made at these points. In addition, the length of the QRS complex is measured. On the basis of these measurements, should a patient be considered "normal" or "abnormal" (that is, a potential heart disease case, or not)? This question might be asked about the tracings for people within a given age–sex group since electrocardiogram patterns are different for different ages and sexes. Rather than look at the measurements individually, it may be easier to combine the data in some way to get a single number that can be used to assign people to the normal or abnormal groups (*503*).

2. A patient is admitted to a hospital with a diagnosis of myocardial infarction. Systolic blood pressure, diastolic blood pressure, heart rate, stroke index, and mean arterial pressure are obtained. Is it possible to predict whether the patient will survive? Can we use these measurements to compute a probability of survival for the patient (*488*)?

3. We have a number of predictors available at five weather stations in an area. These include visibility, height of ceiling, east–west wind component, north–south wind component, total cloud cover, and change in pressure in last 3 hours. On the basis of these measures, we wish to predict what the ceiling will be at an airfield in 2 hours. We must state whether the field will be closed, low instrument, high instrument, low open, or high open. This is an example of a multiple-group discriminant problem (*373*).

4. A geologist has obtained the mean, variance, skewness, and kurtosis of the size of particles deposited in a beach. How can these statistics be used to determine if the beach is wave-laid or aeolian in origin? Are there differences in particle-size distribution (*216*)?

5. In a rural development program, the question arises: What is the best strategy for this area to follow in its development? The problem may also be considered as determining which group the area is most like. Variables might include: distance to nearest city of population over 250,000, number of acres of lakes, percent of land in forest, distance to nearest major

Table 1-1 *Test means by trade*

Trade	*Sample size*	$\bar{x}_1$: *arithmetic*	$\bar{x}_2$: *English*	$\bar{x}_3$: *form relations*
Engineering	404	27.88	98.36	33.60
Building	400	20.65	85.43	31.51
Art	258	15.01	80.31	32.01
Commerce	286	24.38	94.94	26.69

Source: Porebski (*418*).

airport, and so forth. On the basis of these variables, the area might be grouped as catering to recreation users, or attractive to industry (*70*).

6. What are important factors in perinatal death and survival? Can survival be predicted with any degree of accuracy? A study of these questions has been made which uses data from a large number of deliveries (*1a*).

7. An archeologist obtains a skull and wants to know if it belongs to a tribe that inhabited an area 20,000 years ago, or to a successor that lived nearby. On the basis of measurements made on the skull, together with measurements made on a set of skulls from each of the two populations, the assignment may be made.

The common feature of these examples is the goal to assign individuals to a group on the basis of data that are related to the group. In some cases it is desirable to find a subset of the given variables which best perform this assignment.

Consider the following problem. In the last year of secondary school, a student is given three tests. On the basis of scores on an arithmetic test, x_1, an English test, x_2, and a form-relations test, x_3, the student is to be advised on a course of future study [Porebski (*418*)]. The student has four choices available: engineering, building, art, or commerce. A large sample was taken "of entrants to junior technical colleges in the greater London area" and sample means and a pooled covariance matrix were computed. These are given in Tables 1-1 and 1-2.

Table 1-2 *Covariance matrix* $[\mathbf{S} = (s_{ij})]$

Test	x_1	x_2	x_3
x_1	55.58	33.77	11.66
x_2	33.77	360.04	14.53
x_3	11.66	14.53	69.21

Source: Porebski (*418*).

Suppose that an entrant wishes to choose between studying engineering and building. One way to help this person is to see how he compares with the students who have already chosen one of the trades. Suppose that the entrant has scores (x_1, x_2, x_3) and the trades have mean scores $(\bar{x}_{1E}, \bar{x}_{2E}, \bar{x}_{3E})$ and $(\bar{x}_{1B}, \bar{x}_{2B}, \bar{x}_{3B})$. The Euclidian distances between the student's score and the mean trade scores are the sums of squares:

$$D_E^2 = \sum_{i=1}^{3} (x_i - \bar{x}_{iE})^2$$
$$D_B^2 = \sum_{i=1}^{3} (x_i - \bar{x}_{iB})^2 \qquad (1\text{-}1)$$

One might advise the student to enter the trade for which his scores were nearer the mean. This would not be quite right since the presence of correlation can cause different weights to be given to the different tests.

Another distance measure that could be used weighs the observations by a function of the covariances. This is equivalent to a procedure suggested by Fisher (*172*). Let us try to find a linear combination of the observation that gives the greatest amount of squared difference between the two groups relative to the variance within the two groups (that is, which maximizes d^2/v). Suppose that the combination is

$$z = a_1x_1 + a_2x_2 + a_3x_3 \qquad (1\text{-}2)$$

Then the between-group difference is estimated to be

$$d = \bar{z}_E - \bar{z}_B = a_1(\bar{x}_{1E} - \bar{x}_{1B}) + a_2(\bar{x}_{2E} - \bar{x}_{2B}) + a_3(\bar{x}_{3E} - \bar{x}_{3B})$$
$$= a_1(7.23) + a_2(12.93) + a_3(2.09) \qquad (1\text{-}3)$$

The variance of z is

$$v = \sum_i \sum_j a_ia_js_{ij} = a_1^2 \cdot 55.58 + 2a_1a_2 \cdot 33.77 + 2a_1a_3 \cdot 11.66$$
$$+ a_2^2 \cdot 360.04 + 2a_2a_3 \cdot 14.53$$
$$+ a_3^2 \cdot 69.21 \qquad (1\text{-}4)$$

The coefficients a_1, a_2, and a_3 are found by maximizing the quantity d^2/v. This is done by differentiating d^2/v with respect to each coefficient in turn and setting equal to zero. For example,

$$\frac{\partial\, d^2/v}{\partial a_1} = \left(2\frac{vd\, \partial d}{\partial a_1} - \frac{d^2\, \partial v}{\partial a_1}\right) \Big/ v^2 = 0$$

gives

$$2vd(7.23) - 2(d^2)(a_1 \cdot 55.58 + a_2 \cdot 33.77 + a_3 \cdot 11.66) = 0 \qquad (1\text{-}5)$$

Since the coefficients are determined only up to a multiplicative constant (that is, if one set of coefficients maximizes d^2/v, then so does any multiple of it), we can write (1-5) as

$$55.58a_1 + 33.77a_2 + 11.66a_3 = 7.23 \tag{1-6}$$

By differentiating with respect to a_2 and a_3 we get

$$\begin{aligned} 33.77a_1 + 360.04a_2 + 14.53a_3 &= 12.93 \\ 11.66a_1 + 14.53a_2 + 69.21a_3 &= 2.09 \end{aligned} \tag{1-7}$$

The solution to these equations is

$$a_1 = .1136 \qquad a_2 = .0250 \qquad a_3 = .0058$$

The numbers given here differ from Porebski (*418*) because we are using the covariance matrix, whereas Porebski used the within-groups sum-of-squares and products matrix. The function to be used for classifying is

$$D(\mathbf{x}) = .1136x_1 + .0250x_2 + .0058x_3 \tag{1-8}$$

If the chance of being best suited for engineering is the same as the chance of being best suited for building, the best dividing point is midway between the means of the two groups. For the engineers the mean of $D(\mathbf{x})$ is

$$(.1136)(27.88) + (.0250)(98.36) + (.0058)(33.60) = 5.8210$$

and for the builders it is

$$(.1136)(20.65) + (.0250)(85.43) + (.0058)(31.51) = 4.6643$$

The midpoint is $(.5)(5.8210 + 4.6643) \approx 5.24$. Thus the classification rule is: The subject is more like the engineers if $D(\mathbf{x}) > 5.24$ and more like the builders if $D(\mathbf{x}) < 5.24$. Similar analyses could be carried out for all pairs of groups. By sequentially eliminating groups a decision can be reached (in three steps) about which group is most suitable for the individual.

The difference between the group means of the discriminant function is $5.8210 - 4.6643 \approx 1.16$. This quantity is Mahalanobis's distance, which corrects for the effect of correlation. The complete set of distances is given in Table 1-3. It can be seen that engineers and art majors are farthest apart ("most easily separated"); the most similar pair are the builders and art majors. Referring back to the table of means, we see that the builders and art majors have low scores on arithmetic and English and are both in the middle on form relations.

The discriminant function is then used to tell the student which group he most resembles. Because some of the groups overlap considerably, there may be errors in assignment. For example, an engineer may be closer to

Table 1-3 Pairwise Mahalanobis distances

	E	B	A
B	1.16		
A	3.31	0.63	
C	0.70	0.90	2.66

Source: Porebski (*418*).

the mean of the builders than to the mean of the engineers. The Mahalanobis distances may be used to get an approximate notion of the error rates when the observations are approximately normally distributed. If the Mahalanobis distance between groups i and j is denoted by d_{ij}, then the error rate is approximately $1 - \Phi(\sqrt{d_{ij}}/2) = \Phi(-\sqrt{d_{ij}}/2)$. Table 1-4 gives approximate error rates for these data. It can be seen that the groups which are closer together had a higher pairwise error rate than those which are farther apart. A word of caution should be inserted here regarding the use of this estimate of error rates. If the data are not normal, or if the sample sizes are small, this method does not work well. Extreme caution is needed in this situation. We discuss this point in Chapter 2.

A second example concerns advising high school students regarding whether to choose a college-preparatory course or a non-college-preparatory course.

In many high schools, guidance counselors are asked to advise 9th- or 10th-grade students on their choice of high school program. Typically, they will have the student's past records and a battery of standard tests. From these data they recommend a college preparatory or non-college-preparatory program. Lohnes and McIntire (*339a*) reported on the ability to predict satisfactorily on the basis of six tests given by the Educational

Table 1-4 Approximate error rates for Porebski data

	E	B	A
B	.29		
A	.18	.35	
C	.33	.32	.21

Table 1-5 Means and standard deviations of Lohnes–McIntire data

	College prep.		Noncollege prep.	
Test	$\bar{x}$	s	$\bar{x}$	s
SCAT verbal	289.2	11.2	274.6	12.1
SCAT quantitative	302.3	20.0	288.3	19.7
Reading vocabulary	156.4	7.2	147.6	8.2
Reading level	154.7	7.5	146.0	9.0
Reading speed	155.5	8.8	146.0	8.3
Reading comprehension	155.1	8.8	145.5	8.2

Source: Lohnes and McIntire (*335a*).

Testing Service. Means and standard deviations of the observations are given in Table 1-5. There were 404 in the college-preparatory group and 424 in the non-college-preparatory group.

After the discriminant function was computed, the observations were resubstituted into the function and classified. This yields the apparent error rate as given in Table 1-6. This method has a bias, because the same observations are used to calculate the function and to evaluate it. Another way, although it is not always possible, is to have a separate set of data and

Table 1-6 Apparent error rate

	True group		
Assigned to:	*College prep.*	*Noncollege prep.*	*Total*
College prep.	304	112	416
Noncollege prep.	100	312	412
Total	404	424	828

$$P_1 = \frac{100}{404} = .248$$

$$P_2 = \frac{112}{424} = .264$$

$$\bar{P} = \frac{212}{828} = .256$$

Source: Lohnes and McIntire (*335a*).

Table 1-7 *Error rates on separate data set*

Assigned to:	*College prep.*	*Noncollege prep.*	*Total*
College prep.	305	142	447
Noncollege prep.	110	343	453
Total	415	485	900

$$P_1 = \frac{110}{415} = .265$$

$$P_2 = \frac{142}{485} = .293$$

$$\bar{P} = \frac{252}{900} = .280$$

attempt to classify it. Table 1-7 gives the results of such a study. Thus the bias is $.280 - .296 = .024$. In this case it is small, but with smaller samples it can be fairly large.

Another question that should be considered is that of the applicability of the function derived from a statewide sample to a particular school. The method is fairly useless if each school must derive its own function. Lohnes and McIntire tried it in two schools, calculating the optimal rule for the school as well as using the one derived from the statewide data. In the first school, they found an apparent error rate of .21 compared with the error rate for the statewide function when used in that school of .22. In the second school the error rates were .27 and .33, respectively. This difference is uncomfortably large, and one might consider trying to discover why.

It should be noted that the rule assigns students to the groups on the basis of previous self-selected groups. It is possible that there may be considerable initial misclassification in the sense that some of the college-preparatory group should not be in that group, and vice versa. The authors point out that "the criterion does not contain information about the degree of success in the selected curriculum." Indeed, this procedure gives information primarily on the degree of similarity to students who have previously selected these courses of study.

Theory of Discriminant Functions

Suppose that our population consists of two groups, Π_1 and Π_2. We observe a $k \times 1$ vector $\mathbf{x}$ and must assign the individual whose measurements are given by $\mathbf{x}$ to Π_1 or Π_2. We need a rule to assign $\mathbf{x}$ to Π_1 or Π_2. If the parameters of the distributions of $\mathbf{x}$ in Π_1 and Π_2 are known, we may use this knowledge in the construction of an assignment rule. If not, we use samples of size n_1 from Π_1 and n_2 from Π_2 to estimate the parameters. We need a criterion of goodness of classification. Fisher (*172*) has suggested using a linear combination of the observations, and choosing the coefficients so that the ratio of the difference of the means of the linear combination in the two groups to its variance is maximized.

Welch (*567*) suggested that minimizing the total probability of misclassification would be a sensible idea. Von Mises (*545*) suggested minimizing the maximum probability of misclassification in the two groups. Various authors have suggested that different types of misclassifications have different costs (for example, it is more serious to miss an early cancer than to say that a healthy person has cancer) and advocate minimizing the total cost of misclassification. A clear discussion of this appears in Anderson's book (*20*).

In Fisher's approach, let the linear combination be denoted $Y = \boldsymbol{\lambda}'\mathbf{x}$. Then the mean of Y is $\boldsymbol{\lambda}'\boldsymbol{\mu}_1$ in Π_1 and $\boldsymbol{\lambda}'\boldsymbol{\mu}_2$ in Π_2; its variance is $\boldsymbol{\lambda}'\boldsymbol{\Sigma}\boldsymbol{\lambda}$ in either population if we assume that the covariance matrices, $\boldsymbol{\Sigma}_1 = \boldsymbol{\Sigma}_2 = \boldsymbol{\Sigma}$. Then we wish to choose $\boldsymbol{\lambda}$ to maximize

$$\phi = \frac{(\boldsymbol{\lambda}'\boldsymbol{\mu}_1 - \boldsymbol{\lambda}'\boldsymbol{\mu}_2)^2}{\boldsymbol{\lambda}'\boldsymbol{\Sigma}\boldsymbol{\lambda}} \tag{1-9}$$

Differentiating ϕ with respect to $\boldsymbol{\lambda}$ we get

$$\frac{\partial\phi}{\partial\boldsymbol{\lambda}} = \frac{2(\boldsymbol{\mu}_1 - \boldsymbol{\mu}_2)\boldsymbol{\lambda}'\boldsymbol{\Sigma}\boldsymbol{\lambda} - 2\boldsymbol{\Sigma}\boldsymbol{\lambda}(\boldsymbol{\lambda}'\boldsymbol{\mu}_1 - \boldsymbol{\lambda}'\boldsymbol{\mu}_2)}{(\boldsymbol{\lambda}'\boldsymbol{\Sigma}\boldsymbol{\lambda})^2} = \mathbf{0} \tag{1-10}$$

which gives

$$\boldsymbol{\mu}_1 - \boldsymbol{\mu}_2 = \boldsymbol{\Sigma}\boldsymbol{\lambda}\left(\frac{\boldsymbol{\lambda}'\boldsymbol{\mu}_1 - \boldsymbol{\lambda}'\boldsymbol{\mu}_2}{\boldsymbol{\lambda}'\boldsymbol{\Sigma}\boldsymbol{\lambda}}\right) \tag{1-11}$$

Since we use $\boldsymbol{\lambda}$ only to separate the populations, we may multiply $\boldsymbol{\lambda}$ by any constant we desire. Thus $\boldsymbol{\lambda}$ is proportional to $\boldsymbol{\Sigma}^{-1}(\boldsymbol{\mu}_1 - \boldsymbol{\mu}_2)$. If the parameters are not known, it is the usual practice to estimate them by $\bar{\mathbf{x}}_1$, $\bar{\mathbf{x}}_2$, and $\mathbf{S}$. The assignment procedure is to assign an individual to Π_1 if $Y = (\bar{\mathbf{x}}_1 - \bar{\mathbf{x}}_2)'\mathbf{S}^{-1}\mathbf{x}$ is closer to $\bar{Y}_1 = (\bar{\mathbf{x}}_1 - \bar{\mathbf{x}}_2)'\mathbf{S}^{-1}\bar{\mathbf{x}}_1$ then to $\bar{Y}_2$ and to Π_2 otherwise. The midpoint of the interval between $\bar{Y}_1$ and $\bar{Y}_2$ is $(\bar{Y}_1 + \bar{Y}_2)/2 = \frac{1}{2}(\bar{\mathbf{x}}_1 - \bar{\mathbf{x}}_2)'\mathbf{S}^{-1}(\bar{\mathbf{x}}_1 + \bar{\mathbf{x}}_2)$. Y is closer to $\bar{Y}_1$ if

$$|Y - \bar{Y}_1| < |Y - \bar{Y}_2| \tag{1-12}$$

which occurs if

$$Y > \tfrac{1}{2}(\bar{Y}_1 + \bar{Y}_2) \tag{1-13}$$

since $\bar{Y}_1 > \bar{Y}_2$. In Fisher's original paper, he used this method to classify two species of iris. Four measurements were used—sepal length, sepal width, petal length, and petal width—and the classification was excellent. However, petal length or petal width alone would have been adequate to classify the observations. In addition, the assumption of equal covariance matrices for the two species is violated. It may also be noted that the difference between $\bar{Y}_1$ and $\bar{Y}_2$ is

$$\begin{aligned} \bar{Y}_1 - \bar{Y}_2 &= (\bar{\mathbf{x}}_1 - \bar{\mathbf{x}}_2)'\mathbf{S}^{-1}\bar{\mathbf{x}}_1 - (\bar{\mathbf{x}}_1 - \bar{\mathbf{x}}_2)'\mathbf{S}^{-1}\bar{\mathbf{x}}_2 \\ &= (\bar{\mathbf{x}}_1 - \bar{\mathbf{x}}_2)'\mathbf{S}^{-1}(\bar{\mathbf{x}}_1 - \bar{\mathbf{x}}_2) \end{aligned} \tag{1-14}$$

which is Mahalanobis's D^2. Also,

$$\begin{aligned} \text{Var } Y &= \boldsymbol{\lambda}'\mathbf{S}\boldsymbol{\lambda} \\ &= (\bar{\mathbf{x}}_1 - \bar{\mathbf{x}}_2)'\mathbf{S}^{-1}\mathbf{S}\mathbf{S}^{-1}(\bar{\mathbf{x}}_1 - \bar{\mathbf{x}}_2) \\ &= D^2 \end{aligned} \tag{1-15}$$

One can use the distribution of D^2 to test if there are significant differences between the two groups. The variable

$$F = \frac{n_1 n_2 (n_1 + n_2 - k - 1)}{(n_1 + n_2)(n_1 + n_2 - 2)k} D^2 \tag{1-16}$$

where n_1 and n_2 are the sample sizes in Π_1 and Π_2, respectively, and k is the number of variables, has an F distribution with k and $n_1 + n_2 - k - 1$ degrees of freedom.

This method is distribution-free in the sense that it is a reasonable criterion for constructing a linear combination. It will be seen shortly that it is optimal if the observations are multivariate normal. The use of $\frac{1}{2}(\bar{Y}_1 + \bar{Y}_2)$ as a cutoff point can be improved upon if the a priori probabilities of Π_1 and Π_2 are not equal.

Welch (*567*) gave a solution to the problem of minimizing the total probability of misclassification. Let $f_1(\mathbf{x})$ be the density function of $\mathbf{x}$ if it comes from Π_1 and $f_2(\mathbf{x})$ be the density function of $\mathbf{x}$ if it comes from Π_2. Let p_1 be the proportion of Π_1 in the population and $p_2(= 1 - p_1)$ be the proportion of Π_2 in the population. Suppose that we assign $\mathbf{x}$ to Π_1 if $\mathbf{x}$ is in some region R_1 and to Π_2 if $\mathbf{x}$ is in a region R_2. We assume that R_1 and R_2 are mutually exclusive and that their union includes the entire space R. The total probability of misclassification is

$$\begin{aligned} T(R,f) &= p_1 \int_{R_2} f_1(\mathbf{x})\, d\mathbf{x} + p_2 \int_{R_1} f_2(\mathbf{x})\, d\mathbf{x} \\ &= p_1 \left[1 - \int_{R_1} f_1(\mathbf{x})\, d\mathbf{x}\right] + p_2 \int_{R_1} f_2(\mathbf{x})\, d\mathbf{x} \\ &= p_1 + \int_{R_1} [p_2 f_2(\mathbf{x}) - p_1 f_1(\mathbf{x})]\, d\mathbf{x} \end{aligned} \tag{1-17}$$

This quantity is minimized if R_1 is chosen so that $p_2 f_2(\mathbf{x}) - p_1 f_1(\mathbf{x}) < 0$ for all points in R_1. Thus the classification rule is: Assign $\mathbf{x}$ to Π_1 if $f_1(\mathbf{x})/f_2(\mathbf{x}) > p_2/p_1$ and to Π_2 otherwise. This rule minimizes the total probability of misclassification. We define for future reference the probabilities of misclassification within each group:

$$P_1 = \int_{R_2} f_1(\mathbf{x})\, d\mathbf{x}$$

$$P_2 = \int_{R_1} f_2(\mathbf{x})\, d\mathbf{x} \tag{1-18}$$

An important special case is when Π_1 and Π_2 are multivariate normal with means $\boldsymbol{\mu}_1$ and $\boldsymbol{\mu}_2$ and common covariance matrix, $\boldsymbol{\Sigma}$. In Π_i we have

$$f_i(\mathbf{x}) = (2\pi)^{-k/2} \,|\, \boldsymbol{\Sigma} \,|^{-1/2} \exp[-\tfrac{1}{2}(\mathbf{x} - \boldsymbol{\mu}_i)'\boldsymbol{\Sigma}^{-1}(\mathbf{x} - \boldsymbol{\mu}_i)] \tag{1-19}$$

Thus

$$\frac{f_1(\mathbf{x})}{f_2(\mathbf{x})} = \exp[-\tfrac{1}{2}(\mathbf{x} - \boldsymbol{\mu}_1)'\boldsymbol{\Sigma}^{-1}(\mathbf{x} - \boldsymbol{\mu}_1) + \tfrac{1}{2}(\mathbf{x} - \boldsymbol{\mu}_2)'\boldsymbol{\Sigma}^{-1}(\mathbf{x} - \boldsymbol{\mu}_2)]$$

$$= \exp[\mathbf{x}'\boldsymbol{\Sigma}^{-1}(\boldsymbol{\mu}_1 - \boldsymbol{\mu}_2) - \tfrac{1}{2}(\boldsymbol{\mu}_1 + \boldsymbol{\mu}_2)'\boldsymbol{\Sigma}^{-1}(\boldsymbol{\mu}_1 - \boldsymbol{\mu}_2)] \tag{1-20}$$

Taking logarithms we find the optimal rule is assign to Π_1 if

$$D_T(\mathbf{x}) = [\mathbf{x} - \tfrac{1}{2}(\boldsymbol{\mu}_1 + \boldsymbol{\mu}_2)]'\boldsymbol{\Sigma}^{-1}(\boldsymbol{\mu}_1 - \boldsymbol{\mu}_2) > \ln \frac{p_2}{p_1} \tag{1-21}$$

The quantity on the left will be called the true discriminant function $D_T(\mathbf{x})$. Its sample analogue is

$$D_S(\mathbf{x}) = [\mathbf{x} - \tfrac{1}{2}(\bar{\mathbf{x}}_1 + \bar{\mathbf{x}}_2)]'\mathbf{S}^{-1}(\bar{\mathbf{x}}_1 - \bar{\mathbf{x}}_2) \tag{1-22}$$

The coefficients of $\mathbf{x}$ are seen to be identical with Fisher's result for the linear discriminant function.

The function $D_T(\mathbf{x})$ is a linear transformation of $\mathbf{x}$, and it is useful to know its distribution, for this will enable us to calculate the error rates that will occur if we use $D_T(\mathbf{x})$ to assign observations to Π_1 and Π_2. The distribution of $D_T(\mathbf{x})$ is found as follows. Since $\mathbf{x}$ is multivariate normal, $D_T(\mathbf{x})$, being a linear combination of $\mathbf{x}$, is normal. The mean of $D_T(\mathbf{x})$ if $\mathbf{x}$ comes from Π_1 is

$$\begin{aligned} E(D_T(\mathbf{x}) \mid \Pi_1) &= [\boldsymbol{\mu}_1 - \tfrac{1}{2}(\boldsymbol{\mu}_1 + \boldsymbol{\mu}_2)]'\boldsymbol{\Sigma}^{-1}(\boldsymbol{\mu}_1 - \boldsymbol{\mu}_2) \\ &= \tfrac{1}{2}(\boldsymbol{\mu}_1 - \boldsymbol{\mu}_2)'\boldsymbol{\Sigma}^{-1}(\boldsymbol{\mu}_1 - \boldsymbol{\mu}_2) \\ &= \tfrac{1}{2}\delta^2 \end{aligned} \tag{1-23}$$

In Π_2 the mean of $D_T(\mathbf{x})$ is

$$E(D_T(\mathbf{x}) \mid \Pi_2) = [\boldsymbol{\mu}_2 - \tfrac{1}{2}(\boldsymbol{\mu}_1 + \boldsymbol{\mu}_2)]\boldsymbol{\Sigma}^{-1}(\boldsymbol{\mu}_1 - \boldsymbol{\mu}_2) = -\tfrac{1}{2}\delta^2 \quad (1\text{-}24)$$

In either population the variance is

$$\begin{aligned} &E[D_T(\mathbf{x}) - D_T(\boldsymbol{\mu}_i)]^2 \\ &\quad = E[(\mathbf{x} - \boldsymbol{\mu}_i)'\boldsymbol{\Sigma}^{-1}(\boldsymbol{\mu}_1 - \boldsymbol{\mu}_2)]^2 \\ &\quad = E[(\boldsymbol{\mu}_1 - \boldsymbol{\mu}_2)'\boldsymbol{\Sigma}^{-1}(\mathbf{x} - \boldsymbol{\mu}_i)(\mathbf{x} - \boldsymbol{\mu}_i)'\boldsymbol{\Sigma}^{-1}(\boldsymbol{\mu}_1 - \boldsymbol{\mu}_2)] \\ &\quad = (\boldsymbol{\mu}_1 - \boldsymbol{\mu}_2)'\boldsymbol{\Sigma}^{-1}E(\mathbf{x} - \boldsymbol{\mu}_i)(\mathbf{x} - \boldsymbol{\mu}_i)'\boldsymbol{\Sigma}^{-1}(\boldsymbol{\mu}_1 - \boldsymbol{\mu}_2) \\ &\quad = (\boldsymbol{\mu}_1 - \boldsymbol{\mu}_2)'\boldsymbol{\Sigma}^{-1}(\boldsymbol{\mu}_1 - \boldsymbol{\mu}_2) = \delta^2 \end{aligned} \quad (1\text{-}25)$$

The quantity δ^2 is Mahalanobis's distance for known parameters. The probabilities of misclassification are

$$\begin{aligned} P_1 &= \Pr\left[D_T(x) < \ln\frac{1-p_1}{p_1}\right] \\ &= \Pr\left[\frac{D_T(x) - \frac{1}{2}\delta^2}{\delta} < \frac{\ln\dfrac{1-p_1}{p_1} - \dfrac{\delta^2}{2}}{\delta}\right] \\ &= \Phi\left(\frac{\ln\dfrac{1-p_1}{p_1} - \dfrac{\delta^2}{2}}{\delta}\right) \end{aligned} \quad (1\text{-}26)$$

Similarly,

$$P_2 = \Phi\left(-\frac{\ln\dfrac{1-p_1}{p_1} + \dfrac{\delta^2}{2}}{\delta}\right) \quad (1\text{-}27)$$

Table 1-8 gives values of $\ln[(1-p)/p]$ for $0 \le p \le .5$. The important special case $p_1 = .5$ gives $\ln[(1-p_1)/p_1] = 0$, and

$$P_1 = P_2 = \Phi\left(-\frac{\delta}{2}\right) \quad (1\text{-}28)$$

The situation for $D_S(\mathbf{x})$, the sample discriminant, is not so neatly tied up. One can obtain values of means and variances for $D_S(x)$ conditional on the observed $\bar{\mathbf{x}}_1$, $\bar{\mathbf{x}}_2$, and $\mathbf{S}$, but the problem of finding the unconditionality expectation and variance of $D_S(\mathbf{x})$ is more difficult, and the problem of

Table 1-8 $\ln[(1 - p)/p]$, $0 \leq p \leq .5$[a]

	.00	.01	.02	.03	.04	.05	.06	.07	.08	.09
.00	∞	4.595	3.892	3.476	3.178	2.944	2.752	2.587	2.442	2.314
.10	2.197	2.091	1.992	1.901	1.815	1.735	1.658	1.586	1.516	1.450
.20	1.386	1.325	1.266	1.208	1.153	1.099	1.046	.995	.944	.895
.30	.847	.800	.754	.708	.663	.619	.575	.532	.490	.447
.40	.405	.364	.323	.282	.241	.201	.160	.120	.080	.040
.50	0									

[a] Values for $p > .5$ may be obtained by taking the negative of the value for $p' = 1 - p$.

finding the unconditional distribution has not been solved. $D_S(\mathbf{x})$ is conditionally normal given $\bar{\mathbf{x}}_1$, $\bar{\mathbf{x}}_2$, and $\mathbf{S}$.

The means and variances of $D_S(\mathbf{x})$ are

$$\begin{aligned} E(D_S(\mathbf{x}) \mid \Pi_1) &= [\boldsymbol{\mu}_1 - \tfrac{1}{2}(\bar{\mathbf{x}}_1 + \bar{\mathbf{x}}_2)]'\mathbf{S}^{-1}(\bar{\mathbf{x}}_1 - \bar{\mathbf{x}}_2) \\ E(D_S(\mathbf{x}) \mid \Pi_2) &= [\boldsymbol{\mu}_2 - \tfrac{1}{2}(\bar{\mathbf{x}}_1 + \bar{\mathbf{x}}_2)]'\mathbf{S}^{-1}(\bar{\mathbf{x}}_1 - \bar{\mathbf{x}}_2) \\ \operatorname{Var} D_S(\mathbf{x}) &= (\bar{\mathbf{x}}_1 - \bar{\mathbf{x}}_2)'\mathbf{S}^{-1}\boldsymbol{\Sigma}\mathbf{S}^{-1}(\bar{\mathbf{x}}_1 - \bar{\mathbf{x}}_2) \end{aligned} \tag{1-29}$$

One might be tempted to estimate $\boldsymbol{\mu}_1$, $\boldsymbol{\mu}_2$, and $\boldsymbol{\Sigma}$ by $\bar{\mathbf{x}}_1$, $\bar{\mathbf{x}}_2$, and $\mathbf{S}$ to obtain $D^2/2$, $-D^2/2$, and D^2 for the estimates of means and variance, but this leads to problems in estimating error rates (see Chapter 2).

Lachenbruch (*314*) gave unconditional means and variances of $D_S(\mathbf{x})$ for samples of size n_1 and n_2:

$$\begin{aligned} E(D_S(\mathbf{x}) \mid \Pi_1) &= \tfrac{1}{2}C_1\left(\delta^2 - \frac{k(n_2 - n_1)}{n_1 n_2}\right) \\ E(D_S(\mathbf{x}) \mid \Pi_2) &= \tfrac{1}{2}C_1\left(-\delta^2 - \frac{k(n_2 - n_1)}{n_1 n_2}\right) \end{aligned} \tag{1-30}$$

where

$$C_1 = \frac{n_1 + n_2 - 2}{n_1 + n_2 - k - 3}$$

and

$$\operatorname{Var} D_S(\mathbf{x}) = C_2\left(\delta^2 + \frac{k(n_1 + n_2)}{n_1 n_2}\right) \tag{1-31}$$

where

$$C_2 = \frac{(n_1 + n_2 - 3)(n_1 + n_2 - 2)^2}{(n_1 + n_2 - k - 2)(n_1 + n_2 - k - 3)(n_1 + n_2 - k - 5)}$$

Although $D_S(\mathbf{x})$ is conditionally distributed normally, the unconditional distribution is not normal. Many statisticians have studied this problem, including Wald (*550*), Anderson (*18, 23, 24*), Sitgreaves (*489*), Kabe (*281*), and Okamoto (*398*). Okamoto gave an asymptotic expansion for the distribution of $D_S(\mathbf{x})$.

Bayes theorem approach

Another approach that leads to the above classification rules is to assign $\mathbf{x}$ to the group with the largest posterior probability. By definition, the conditional density of $\mathbf{x}$ given Π_i is $f_i(\mathbf{x})$. Since the a priori probability of Π_i is p_i, the posterior probability of Π_i by Bayes theorem is

$$\Pr(\Pi_i \mid \mathbf{x}) = \frac{\Pr(\Pi_i, \mathbf{x})}{\Pr(\mathbf{x})} = \frac{p_i f_i(\mathbf{x})}{p_1 f_1(\mathbf{x}) + p_2 f_2(\mathbf{x})} \qquad i = 1, 2 \quad \text{(1-32)}$$

If we assign an observation to Π_1 when $\Pr(\Pi_1 \mid \mathbf{x}) > \Pr(\Pi_2 \mid \mathbf{x})$, this is equivalent to the rule that minimizes the total probability of misclassification. When estimating the risk of belonging to Π_1 (for example, having a disease, or dying), these posterior probabilities are useful. This is discussed further in Chapter 6.

Unequal costs of misclassification

An alternative procedure is based on minimizing the total cost of misclassification. In many cases it is more serious to make one kind of error than the other. For example, failure to detect an early cancer is costlier than stating that a patient has cancer and then discovering otherwise. Let C_1 be the cost of misclassifying a member of Π_1 and C_2 be the cost of misclassifying Π_2. Then we wish to find R_1 and R_2 to minimize

$$\begin{aligned} T_1 &= C_1 p_1 \int_{R_2} f_1(\mathbf{x})\, d\mathbf{x} + C_2 p_2 \int_{R_1} f_2(\mathbf{x})\, d\mathbf{x} \\ &= C_1 p_1 \left[1 - \int_{R_1} f_1(\mathbf{x})\, d\mathbf{x}\right] + C_2 p_2 \int_{R_1} f_2(\mathbf{x})\, d\mathbf{x} \\ &= C_1 p_1 + \int_{R_1} [C_2 p_2 f_2(\mathbf{x}) - C_1 p_1 f_1(\mathbf{x})]\, d\mathbf{x} \qquad \text{(1-33)} \end{aligned}$$

which is minimized if R_1 is chosen so that $C_2p_2f_2(\mathbf{x}) < C_1p_1f_1(\mathbf{x})$ in R_1. This is equivalent to choosing Π_1 if

$$\frac{f_1(\mathbf{x})}{f_2(\mathbf{x}_2)} > \frac{C_2p_2}{C_1p_1} \tag{1-34}$$

One might also define weights

$$w_1 = \frac{C_1p_1}{C_1p_1 + C_2p_2}$$

$$w_2 = \frac{C_2p_2}{C_1p_1 + C_2p_2}$$

and minimize

$$T_1 = w_1P_1 + w_2P_2$$

Thus the problem is equivalent to minimizing the total error rate for some a priori "probabilities," w_i.

The main problem with minimizing the cost of misclassification is the difficulty in specifying these costs. The specification of costs is usually done by the user rather than the statistician, and most users are not able to do so. In reality all that is needed is the ratio of costs, C_2/C_1, but even this is hard to get. A constrained discrimination method proposed by Anderson (*13*) to alleviate this problem will be discussed later.

Minimax rule

In some situations we wish to protect ourselves from a rule that does very badly on one population. In this case, we set up a rule that minimizes the maximum probability of misclassification. This procedure is known as a minimax procedure. We desire to set up regions R_1 and R_2 such that

$$P_1 = \int_{R_2} f_1(\mathbf{x})\,d\mathbf{x} = \int_{R_1} f_2(\mathbf{x})\,d\mathbf{x} = P_2 \tag{1-35}$$

and are a minimum. Now

$$\int_{R_2} f_1(\mathbf{x})\,d\mathbf{x} = 1 - \int_{R_1} f_1(\mathbf{x})\,d\mathbf{x} = \int_{R_1} f_2(\mathbf{x})\,d\mathbf{x}$$

or

$$\int_{R_1} [f_1(\mathbf{x}) + f_2(\mathbf{x})]\,d\mathbf{x} = 1$$

Thus we wish to minimize (using LaGrange multipliers)

$$\int_{R_1} f_2(\mathbf{x}) - \alpha[f_1(\mathbf{x}) + f_2(\mathbf{x})]\, d\mathbf{x} + \alpha$$

$$\propto \int_{R_1} [\beta f_2(\mathbf{x}) - f_1(\mathbf{x})]\, d\mathbf{x} + C \tag{1-36}$$

This is minimized if $\beta f_2(\mathbf{x}) - f_1(\mathbf{x})$ is negative throughout R_1. Thus the minimax rule has the form: Assign to Π_1 if

$$\frac{f_1(\mathbf{x})}{f_2(\mathbf{x})} > \beta \tag{1-37}$$

and to Π_2 otherwise. The value of β is determined by the distributions involved. For the normal case, with equal covariances, $\beta = 1$. If $f_i(\mathbf{x}) = \lambda_i e^{-\lambda_i x}$, $\lambda_1 < \lambda_2$, then β is the solution of

$$1 - e^{-\lambda_2 \beta} = e^{-\lambda_1 \beta} \tag{1-38}$$

The above discussion closely follows Kendall (*288a*).

Sample size

In practice, we do not have population parameters available, but using their maximum likelihood estimates has been shown to be a satisfactory procedure. Well-known theorems show that $D_S(\mathbf{x})$ is a consistent estimate

Table 1-9 *Sample size needed to be within .05 of optimum error rate*[a]

k	δ	n
2	1	9
	2	8
	3	7
10	1	35
	2	27
	3	20
20	1	67
	2	51
	3	38

Source: Lachenbruch (*314*).

[a] Optimum values: $P_1 = .309$ if $\delta = 1$, .159 if $\delta = 2$, .067 if $\delta = 3$.

of $D_T(\mathbf{x})$. Hoel and Peterson (*244*) and Glick (*209*) have studied the problem of optimality of these procedures. In general, the discriminant function performs fairly well with samples of moderate size. Table 1-9 gives the sample size needed for various values of δ and k if we wish to be within .05 of the optimum error rate. This assumes normality, $n_1 = n_2$, and equal a priori probabilities for Π_1 and Π_2. We see that about $2k$ observations are needed if δ is large, $3.5k$ if δ is small. If we wish to be within .01 of the true value, the number of observations is about $10k$ for $\delta = 3$ and $20k$ for $\delta = 1$. The number of observations needed depends on the tolerance desired and the size of the between-group distances.

Regression analogy

There is an interesting parallel between the linear discriminant function and the multiple linear regression of the predictor variable on a dummy-variable indicator of group membership [Cramer (*133*), Kendall (*288a*), Fisher (*172*)]. To this end let

$$y_i = \frac{n_2}{n_1 + n_2} \qquad \text{if } \mathbf{x}_i \text{ is a member of } \Pi_1$$

$$y_i = -\frac{n_1}{n_1 + n_2} \qquad \text{if } \mathbf{x}_i \text{ is a member of } \Pi_2$$

This gives $\bar{\mathbf{y}} = 0$. We wish to find parameters $\boldsymbol{\lambda}$ which best fit the model

$$E(y_i) = \boldsymbol{\lambda}'(\mathbf{x}_i - \bar{\mathbf{x}}) \tag{1-39}$$

Note first that

$$\bar{\mathbf{x}} = \frac{n_1\bar{\mathbf{x}}_1 + n_2\bar{\mathbf{x}}_2}{n_1 + n_2} \tag{1-40}$$

and

$$\sum_{i=1}^{n_1+n_2} y_i(\mathbf{x}_i - \bar{\mathbf{x}}) = \frac{n_2}{n_1 + n_2} n_1(\bar{\mathbf{x}}_1 - \bar{\mathbf{x}}) - \frac{n_1}{n_1 + n_2} n_2(\bar{\mathbf{x}}_2 - \bar{\mathbf{x}})$$

$$= \frac{n_1 n_2}{n_1 + n_2} (\bar{\mathbf{x}}_1 - \bar{\mathbf{x}}_2) \tag{1-41}$$

Next, we find

$$\sum_{i=1}^{n_1+n_2} (\mathbf{x}_i - \bar{\mathbf{x}})(\mathbf{x}_i - \bar{\mathbf{x}})' = \sum_{i=1}^{n_1} (\mathbf{x}_i - \bar{\mathbf{x}})(\mathbf{x}_i - \bar{\mathbf{x}})'$$

$$+ \sum_{i=n_1+1}^{n_2} (\mathbf{x}_i - \bar{\mathbf{x}})(\mathbf{x}_i - \bar{\mathbf{x}})' \tag{1-42}$$

The two terms may be written as

$$\sum_{i=1}^{n_1} (\mathbf{x}_i - \bar{\mathbf{x}}_1)(\mathbf{x}_i - \bar{\mathbf{x}}_1)' + n_1(\bar{\mathbf{x}}_1 - \bar{\mathbf{x}})(\bar{\mathbf{x}}_1 - \bar{\mathbf{x}})'$$

and

$$\sum_{i=n_1+x}^{n_2} (\mathbf{x}_i - \bar{\mathbf{x}}_2)(\mathbf{x}_i - \bar{\mathbf{x}}_2)' + n_2(\bar{\mathbf{x}}_2 - \bar{\mathbf{x}})(\bar{\mathbf{x}}_2 - \bar{\mathbf{x}}) \qquad (1\text{-}43)$$

Thus we have

$$\Sigma(\mathbf{x}_i - \bar{\mathbf{x}})(\mathbf{x}_i - \bar{\mathbf{x}})' = (n_1 + n_2 - 2)\mathbf{S} + \frac{n_1 n_2}{n_1 + n_2}(\bar{\mathbf{x}}_1 - \bar{\mathbf{x}}_2)(\bar{\mathbf{x}}_1 - \bar{\mathbf{x}}_2)' \qquad (1\text{-}44)$$

The normal equations for the regression are

$$\left[(n_1 + n_2 - 2)\mathbf{S} + \frac{n_1 n_2}{n_1 + n_2}(\bar{\mathbf{x}}_1 - \bar{\mathbf{x}}_2)(\bar{\mathbf{x}}_1 - \bar{\mathbf{x}}_2)'\right]\boldsymbol{\lambda} = \frac{n_1 + n}{n_1 + n_2}(\bar{\mathbf{x}}_1 - \bar{\mathbf{x}}_2) \qquad (1\text{-}45)$$

Letting $A = (\bar{\mathbf{x}}_1 - \bar{\mathbf{x}}_2)'\boldsymbol{\lambda}$, we have

$$(n_1 + n_2 - 2)\mathbf{S}\boldsymbol{\lambda} = \frac{n_1 n_2}{n_1 + n_2}(\bar{\mathbf{x}}_1 - \bar{\mathbf{x}}_2)(1 - A) \qquad (1\text{-}46)$$

Hence $\boldsymbol{\lambda}$ is proportional to $\mathbf{S}^{-1}(\bar{\mathbf{x}}_1 - \bar{\mathbf{x}}_2)$, the discriminant function coefficients obtained earlier.

This immediately suggests an analogy with the analysis of variance. For this regression the appropriate ANOVA table is as given in Table 1-10. The usual F test of homogeneity may be applied here. If significant, this is evidence that there are between-group differences. It does not mean that

Table 1-10 ANOVA table

Source	*Sum of squares*	*df*
Due to regression	$\frac{n_1 n_2}{n_1 + n_2}\boldsymbol{\lambda}'(\bar{\mathbf{x}}_1 - \bar{\mathbf{x}}_2)$	k
About regression	$\frac{n_1 n_2}{n_1 + n_2}[1 - \boldsymbol{\lambda}'(\bar{\mathbf{x}}_1 - \bar{\mathbf{x}}_2)]$	$n_1 + n_2 - k - 1$
Total	$\frac{n_1 n_2}{n_1 + n_2}$	$n_1 + n_2 - 1$

the discriminant function will be useful for assigning individuals to groups, because significance may be due to large numbers of observations. This is equivalent to performing the usual tests of significance on D^2.

There is also an interesting relationship between R^2, the multiple correlation coefficient, and D^2. If a discriminant function is obtained by using the regression technique, one may calculate a value of R^2 from the ANOVA table in the usual manner. Because the test of significance is identical with the one based on D^2, it is easy to show that

$$D^2 = \frac{R^2}{1 - R^2} \frac{(n_1 + n_2)(n_1 + n_2 - 2)}{n_1 n_2} \tag{1-47}$$

This holds only if the dependent variable is dichotomous.

Computer programs

Many computer programs are available to perform linear discriminant analyses. Perhaps the most widely used package is BMD (*151*), which has three discriminant analysis programs, BMD 04M, BMD 05M, and BMD 07M. BMD 04M computes a discriminant function for two groups using specified subsets of variables. The output includes group means, covariance matrix, coefficients of the function, Mahalanobis's D^2, and the ordered values of the discriminant function for each case, together with group identification numbers. BMD 05M performs a multiple-group discriminant analysis for up to five groups. Output includes means, covariance matrix, Mahalanobis's generalized D^2, coefficients, evaluation of the discriminant function for each case, and a classification matrix. It is assumed that the a priori probabilities are the same for each group, which can be a rather serious limitation. BMD 07M performs a stepwise discriminant analysis on up to 80 groups. The variable to enter or be deleted is selected on the basis of one of three criteria at the user's option. This program also has the option of specifying prior probabilities. Output includes the group means and pooled covariance matrix, classification matrices at specified steps, and posterior probabilities of coming from each group, among others. If this program is used for a two-group problem, the usual coefficients are found by subtracting the coefficients printed here. A special feature of this program is the availability of plotting canonical variates (this is useful in multiple-group problems). One word of caution should be inserted here. Stepwise programs are hazardous to the unwary. If the user allows the program to select the 10 "best" variables, there is a strong possibility that some of these "best" variables are mere noise. Certainly no variable that enters *after* a nonsignificant entry should be considered, and I believe that usually three to five variables is the maximum number that can be safely selected in this way.

Other packages have similar programs. The user should become familiar with the features available in the programs. In particular, one of the most useful features is the ability to specify a priori probabilities.

Quadratic discrimination

The equal covariance assumption is rarely satisfied, although in some cases the two matrices are close enough that it makes little or no difference in the results to assume equality. When the covariance matrices are quite different and normality holds, the optimal rule is: Assign to Π_1 if

$$\begin{aligned} Q(\mathbf{x}) &= \ln \frac{f_1(\mathbf{x})}{f_2(\mathbf{x})} > \ln \frac{1 - p_1}{p_1} \\ &= \tfrac{1}{2} \ln \frac{|\mathbf{\Sigma}_2|}{|\mathbf{\Sigma}_1|} - \tfrac{1}{2}(\mathbf{x} - \boldsymbol{\mu}_1)'\mathbf{\Sigma}_1^{-1}(\mathbf{x} - \boldsymbol{\mu}_1) + \tfrac{1}{2}(\mathbf{x} - \boldsymbol{\mu}_2)'\mathbf{\Sigma}_2^{-1}(\mathbf{x} - \boldsymbol{\mu}_2) \\ &= C_0 - \tfrac{1}{2}[\mathbf{x}'(\mathbf{\Sigma}_1^{-1} - \mathbf{\Sigma}_2^{-1})\mathbf{x} - 2\mathbf{x}'(\mathbf{\Sigma}_1^{-1}\boldsymbol{\mu}_1 - \mathbf{\Sigma}_2^{-1}\boldsymbol{\mu}_2)] \end{aligned} \tag{1-48}$$

In this case one has a quadratic discriminant function since $\mathbf{\Sigma}_1^{-1} - \mathbf{\Sigma}_2^{-1}$ does not vanish. In practice, deviations from normality tend to affect this function rather seriously. Although in theory this is a fine procedure, it is not robust to nonnormality, particularly if the distribution has longer tails than the normal.

Bartlett and Please (*42*) used data originally given by Stocks (*514*) on twins to study the usefulness of the quadratic approach in discriminating monozygotic and dizygotic twins. A series of measures were taken on each child, and the differences between twins were used as the discriminators. As either twin may be chosen first, the differences have mean zero in both groups. The only way to discriminate is by using the differences in covariance matrices. They consider two special cases. In the first case, it is assumed that the variables are independent, and $\mathbf{\Sigma}_i = \sigma_i^2\mathbf{I}$; in the second, they assumed that there were equal correlations and

$$\mathbf{\Sigma}_i = \sigma_i^2 \begin{pmatrix} 1 & \rho_i & \cdots & & \rho_i \\ \rho_i & 1 & & & \rho_i \\ & & \cdot & & \\ & & & \cdot & \\ & & & & \cdot \\ \rho_i & & & \rho_i & 1 \end{pmatrix} = \sigma_i^2\mathbf{R}_i \tag{1-49}$$

Now

$$\ln \frac{f_1(\mathbf{x})}{f_2(\mathbf{x})} = \tfrac{1}{2}[\mathbf{x}'(\mathbf{\Sigma}_2^{-1} - \mathbf{\Sigma}_1^{-1})\mathbf{x}] + \tfrac{1}{2} \ln \frac{|\mathbf{\Sigma}_2|}{|\mathbf{\Sigma}_1|} \tag{1-50}$$

If $\mathbf{\Sigma}_i = \sigma_i^2\mathbf{I}$, then

$$\ln\frac{f_1(\mathbf{x})}{f_2(\mathbf{x})} = \tfrac{1}{2}\mathbf{x}'\mathbf{x}\left(\frac{1}{\sigma_2^2} - \frac{1}{\sigma_1^2}\right) + \frac{k}{2}\ln\frac{\sigma_2^2}{\sigma_1^2} \tag{1-51}$$

If $\mathbf{\Sigma}_i = \sigma_i^2\mathbf{R}_i$, then

$$\mathbf{\Sigma}_i^{-1} = \frac{1}{\sigma_i^2}\left[\frac{\mathbf{I}}{1-\rho_i} - \frac{\rho_i}{1-\rho_i}\frac{\mathbf{E}}{1+(k-1)\rho_i}\right] \tag{1-52}$$

where $\mathbf{E} = (e_{ij})$ and $e_{ij} = 1$ for all i, j. Now $\mathbf{x}'\mathbf{I}\mathbf{x} = \mathbf{x}'\mathbf{x} = \sum_{j=1}^{k} x_j^2 = Z_1$ and $\mathbf{x}'\mathbf{E}\mathbf{x} = (\Sigma x_j)^2 = Z_2$. Thus (1-50) becomes

$$-2\ln\frac{f_1(\mathbf{x})}{f_2(\mathbf{x})} = c_1Z_1 - c_2Z_2 + c_3 \tag{1-53}$$

where

$$c_1 = \frac{1}{\sigma_1^2(1-\rho_1)} - \frac{1}{\sigma_2^2(1-\rho_2)},$$

$$c_2 = \frac{\rho_1}{\sigma_1^2(1-\rho_1)}\frac{1}{1+(k-1)\rho_1} - \frac{\rho_2}{\sigma_2^2(1-\rho_2)}\frac{1}{1+(k-1)\rho_2}$$

and the rule is: Assign to Π_1 if

$$c_1Z_1 - c_2Z_2 + c_3 < -2\ln\frac{1-p_1}{p_1} \tag{1-54}$$

If $\rho_1 = \rho_2 = 0$, $c_2 = 0$ and the assignment is based entirely on Z_1 (this is not the only case in which $c_2 = 0$). If $\sigma_1^2(1-\rho_1) = \sigma_2^2(1-\rho_2)$, $c_1 = 0$ and the classification is based entirely on Z_2. Z_1 has been interpreted as indicative of the "shape" of the observation and Z_2 as the "size" of the observation.

Using the Stocks data and assuming that $\rho_1 = \rho_2$, they calculated a discriminant function based on 10 variables, using 15 twin pairs in each group. This was done for males and females separately. They found that among males 4 monozygotic and 7 dizygotic twins were misallocated, and among females no monozygotic and 2 dizygotic twins were misallocated. These results must be treated cautiously because of the small sample size used.

These data were also studied by Desu and Geisser (*149*) in a Bayesian context. They also extended the methods to multiple birth discrimination.

It may be noted that other distributions than the normal lead to quadratic functions. Cooper (*120*) has given two multivariate generalizations of Pearson type II and type VII distributions which possess this property. The type II distribution is defined as

$$f_i(\mathbf{x}) = \frac{\Gamma(m + \frac{1}{2}k + 1)}{(\pi)^{k/2}\Gamma(m + 1)} |\mathbf{W}_i|^{1/2}[1 - (\mathbf{x} - \boldsymbol{\mu}_i)'\mathbf{W}_i(\mathbf{x} - \boldsymbol{\mu}_i)]^m \quad (1\text{-}55)$$

if

$$(\mathbf{x} - \boldsymbol{\mu}_i)'\mathbf{W}_i(\mathbf{x} - \boldsymbol{\mu}_i) \leq 1$$

$$= 0 \qquad \text{otherwise}$$

where $m \geq 0$ and is assumed the same in Π_1 and Π_2. Then

$$\frac{f_1(\mathbf{x})}{f_2(\mathbf{x})} = \frac{|\mathbf{W}_1|^{1/2}}{|\mathbf{W}_2|^{1/2}} \left[\frac{1 - (\mathbf{x} - \boldsymbol{\mu}_1)'\mathbf{W}_1(\mathbf{x} - \boldsymbol{\mu}_1)}{1 - (\mathbf{x} - \boldsymbol{\mu}_2)'\mathbf{W}_2(\mathbf{x} - \boldsymbol{\mu}_2)}\right]^m \quad (1\text{-}56)$$

The classification rule may be written: Assign to Π_1 if

$$\left[\frac{f_1(\mathbf{x})}{f_2(\mathbf{x})}\right]^{1/m} = \left[\frac{|\mathbf{W}_1|}{|\mathbf{W}_2|}\right]^{1/2m} \left[\frac{1 - (\mathbf{x} - \boldsymbol{\mu}_1)'\mathbf{W}_1(\mathbf{x} - \boldsymbol{\mu}_1)}{1 - (\mathbf{x} - \boldsymbol{\mu}_2)'\mathbf{W}_2(\mathbf{x} - \boldsymbol{\mu}_2)}\right] > \left(\frac{1 - p_1}{p_1}\right)^{1/m} \quad (1\text{-}57)$$

or, equivalently: Assign to Π_1 if

$$C(\mathbf{x} - \boldsymbol{\mu}_2)'\mathbf{W}_2(\mathbf{x} - \boldsymbol{\mu}_2) - (\mathbf{x} - \boldsymbol{\mu}_1)'\mathbf{W}_1(\mathbf{x} - \boldsymbol{\mu}_1) > C - 1 \quad (1\text{-}58)$$

If $\mathbf{W}_1 = \mathbf{W}_2$ and $p_1 = p_2 = \frac{1}{2}$, $C = 1$ and the rule becomes: Assign to Π_1 if

$$\mathbf{x}'\mathbf{W}(\boldsymbol{\mu}_1 - \boldsymbol{\mu}_2) - \tfrac{1}{2}(\boldsymbol{\mu}_1 + \boldsymbol{\mu}_2)'\mathbf{W}(\boldsymbol{\mu}_1 - \boldsymbol{\mu}_2) > 0 \quad (1\text{-}59)$$

Thus the linear or quadratic functions arise as the classification rule for the multivariate type II distributions. The covariance matrix of $\mathbf{x}$ is

$$\boldsymbol{\Sigma}_i = \frac{1}{2m + k + 2} \mathbf{W}_i^{-1} \quad (1\text{-}60)$$

The multivariate type VII distribution is defined as

$$f_i(\mathbf{x}) = \frac{\Gamma(m)}{\pi^{k/2}\Gamma(m - k/2)} |\mathbf{W}_i|^{1/2}[1 + (\mathbf{x} - \boldsymbol{\mu}_i)'\mathbf{W}_i(\mathbf{x} - \boldsymbol{\mu}_i)]^{-m} \quad (1\text{-}61)$$

where $2m > k$. In this case

$$\frac{f_1(\mathbf{x})}{f_2(\mathbf{x})} = \frac{|\mathbf{W}_1|^{1/2}}{|\mathbf{W}_2|^{1/2}} \left[\frac{1 + (\mathbf{x} - \boldsymbol{\mu}_2)'\mathbf{W}_2(\mathbf{x} - \boldsymbol{\mu}_2)}{1 + (\mathbf{x} - \boldsymbol{\mu}_1)'\mathbf{W}_1(\mathbf{x} - \boldsymbol{\mu}_1)}\right]^m \quad (1\text{-}62)$$

so the assignment rule has the form

$$-C(\mathbf{x} - \boldsymbol{\mu}_1)'\mathbf{W}_1(\mathbf{x} - \boldsymbol{\mu}_1) + (\mathbf{x} - \boldsymbol{\mu}_2)'\mathbf{W}_2(\mathbf{x} - \boldsymbol{\mu}_2) > C - 1 \quad (1\text{-}63)$$

Again the linear discriminant results if $\mathbf{W}_1 = \mathbf{W}_2$ and $p_1 = p_2 = \frac{1}{2}$. It is interesting to note that even if $\mathbf{W}_1 = \mathbf{W}_2$, the linear discriminant is not the optimal classification rule unless $p_1 = p_2 = \frac{1}{2}$ also. For the type VII distribution the covariance matrix is

$$\mathbf{\Sigma}_i = (2m - k - 2)\mathbf{W}_i^{-1} \tag{1-64}$$

Problems

1. Show that if $\mathbf{x} \sim N(\boldsymbol{\mu}, \mathbf{\Sigma})$ and $\mathbf{y} = \mathbf{Cx}$, where $\mathbf{C}$ is a matrix, then $\mathbf{y} \sim N(\mathbf{C}\boldsymbol{\mu}_1, \mathbf{C\Sigma C}')$.
2. Mahalanobis δ^2 is defined as $\delta^2 = (\boldsymbol{\mu}_1 - \boldsymbol{\mu}_2)'\mathbf{\Sigma}^{-1}(\boldsymbol{\mu}_1 - \boldsymbol{\mu}_2)$ when $\mathbf{x}$ is $N(\boldsymbol{\mu}_1, \mathbf{\Sigma})$ in Π_1 and $\mathbf{x}$ is $N(\boldsymbol{\mu}_2, \mathbf{\Sigma})$ in Π_2. Show that δ^2 is invariant under nonsingular linear transformations (that is, if $y = \mathbf{Cx}$ and $\mathbf{C}$ is nonsingular, then $\delta_y^2 = \delta_x^2$).
3. Let $\boldsymbol{\lambda}'\mathbf{x}$ be a linear combination of $\mathbf{x}$, and let $\mathbf{d} = (\bar{\mathbf{x}}_1 - \bar{\mathbf{x}}_2)$ and $\mathbf{S}$ be the covariance matrix of $\mathbf{x}$. Show that if $\boldsymbol{\lambda}$ satisfies

$$\frac{\boldsymbol{\lambda}'\mathbf{dd}'\boldsymbol{\lambda}}{\boldsymbol{\lambda}'\mathbf{S}\boldsymbol{\lambda}} = \eta$$

   then so does $a\boldsymbol{\lambda}$, where a is any constant. This shows that the discriminant coefficients are determined up to a constant multiple.
4. Verify the solutions of (1-6) and (1-7):
   (a) By substituting the values.
   (b) By solving the equations yourself.
   If you get different answers from the ones in the book it is because you carried more or fewer significant figures.
5. Mahalanobis D^2 is the sample analogue of δ^2 and is given by $D^2 = (\bar{\mathbf{x}}_1 - \bar{\mathbf{x}}_2)'\mathbf{S}^{-1}(\bar{\mathbf{x}}_1 - \bar{\mathbf{x}}_2)$. Use the data in Tables 1-1 and 1-2 to verify the entries of Table 1-3.
6. Find the sample discriminant function to assign to engineering (Π_1) or art (Π_2), and estimate the error rate if $p_1 = .60$.
7. Suppose that you found error rates of $P_1 = .05$ and $P_2 = .10$ for a problem in which $D_T(\mathbf{x})$ was known. Find the value of the cutoff point and the value of δ^2.
8. Suppose that $k = 3$, and $\mathbf{x} \sim N(\mathbf{0}, \mathbf{I})$ in Π_1 and $\mathbf{x} \sim N(\mathbf{0}, 4\mathbf{I})$i n Π_2. Find the optimal rule for assigning $\mathbf{x}$ to Π_1 or Π_2 for general p_1. Find P_1 and P_2 if $p_1 = \frac{1}{2}$. What can you say about error rates if the covariance matrices are not proportional?

Projects

1. Obtain a set of data on two populations and perform a discriminant analysis on it using a packaged program such as BMD 04M. Data are available in Fisher (*172*) and Lubischew (*338*). Data suitable for quadratic discrimination are found in Stocks (*514*).

2. Write a program to calculate the linear discriminant function. Include in your output the means and standard deviations, Mahalanobis D^2 and discriminant coefficients. You should have the ability to modify the cutoff point and to estimate the error rates given by (1-26) and (1-27). Be sure to test your program for accuracy.
3. Write a program to calculate a quadratic discriminant function. Note that you will need to store two covariance matrices.

2 *Evaluating a discriminant function*

After a discriminant function has been calculated, its performance should be evaluated. Three questions are of major importance:

1. Are the observed between-group differences real? That is, are the differences we observe statistically significant? This determines if there is any hope of classifying future observations using the given variables. If not, then we might as well try to find better variables.

2. If the differences are greater than would be expected by chance, are all of the variables needed? We wish, in general to reduce the number of variables in the discriminant function. Alternatively we may restate the question as: Is a subset of the variables sufficient for discrimination?

3. How will the discriminant function perform on future samples? This involves estimating the error rates of the given discriminant function. Various techniques of estimating error rates have been proposed; some depend on normality, others do not.

Tests of between-group differences

The first problem may be solved in one of two equivalent ways introduced in Chapter 1. Mahalanobis's D^2 may be calculated as

$$D^2 = (\bar{\mathbf{x}}_1 - \bar{\mathbf{x}}_2)'\mathbf{S}^{-1}(\bar{\mathbf{x}}_1 - \bar{\mathbf{x}}_2) \tag{2-1}$$

It is known from (1-16) that

$$F = \frac{n_1 n_2}{n_1 + n_2} \frac{(n_1 + n_2 - k - 1)}{(n_1 + n_2 - 2)k} D^2 \tag{2-2}$$

Calculation of F is equivalent to performing a T^2 test since $T^2 = (n_1 n_2/n_1 + n_2)D^2$ [see Morrison (*378*)]. The resulting F statistic has k and $n_1 + n_2 - k - 1$ df. The second method is to use the analysis of variance. Recall that the regression approach yields discriminant coefficients which are proportional to $\mathbf{S}^{-1}(\bar{\mathbf{x}}_1 - \bar{\mathbf{x}}_2)$. Following Kendall (*289*) we have, from (1-46),

$$(n_1 + n_2 - 2)\mathbf{S}\lambda = \frac{n_1 n_2}{n_1 + n_2}(\bar{\mathbf{x}}_1 - \bar{\mathbf{x}}_2)(1 - A)$$

and hence [multiplying on the left by $(\bar{\mathbf{x}}_1 - \bar{\mathbf{x}}_2)'\mathbf{S}^{-1}$]

$$A = \frac{n_1 n_2}{(n_1 + n_2)(n_1 + n_2 - 2)}(\bar{\mathbf{x}}_1 - \bar{\mathbf{x}}_2)'\mathbf{S}^{-1}(\bar{\mathbf{x}}_1 - \bar{\mathbf{x}}_2)(1 - A) \qquad (2\text{-}3)$$

which gives

$$A = \frac{\dfrac{n_1 n_2}{(n_1 + n_2)(n_1 + n_2 - 2)} D^2}{1 + \dfrac{n_1 n_2}{(n_1 + n_2)(n_1 + n_2 - 2)} D^2} \qquad (2\text{-}4)$$

Let $C = n_1 n_2/(n_1 + n_2)(n_1 + n_2 - 2)$. Then $A = CD^2/1 + CD^2$ and

$$\lambda = C\mathbf{S}^{-1}(\bar{\mathbf{x}}_1 - \bar{\mathbf{x}}_2)\left(1 - \frac{CD^2}{1 + CD^2}\right) = \frac{C\mathbf{S}^{-1}(\bar{\mathbf{x}}_1 - \bar{\mathbf{x}}_2)}{1 + CD^2}$$

From the ANOVA table we find the F ratio:

$$F = \frac{\lambda'(\bar{\mathbf{x}}_1 - \bar{\mathbf{x}}_2)}{1 - \lambda'(\bar{\mathbf{x}}_1 - \bar{\mathbf{x}}_2)} \frac{n_1 + n_2 - k - 1}{k}$$

$$= \frac{\dfrac{C}{1 + CD^2}(\bar{\mathbf{x}}_1 - \bar{\mathbf{x}}_2)'\mathbf{S}^{-1}(\bar{\mathbf{x}}_1 - \bar{\mathbf{x}}_2)}{1 - \dfrac{C(\bar{\mathbf{x}}_1 - \bar{\mathbf{x}}_2)'\mathbf{S}^{-1}(\bar{\mathbf{x}}_1 - \bar{\mathbf{x}}_2)}{1 + CD^2}} \frac{n_1 + n_2 - k - 1}{k}$$

$$= \frac{\dfrac{CD^2}{1 + CD^2}}{\dfrac{1}{1 + CD^2}} \frac{n_1 + n_2 - k - 1}{k} \qquad (2\text{-}5)$$

$$F = \frac{n_1 n_2}{(n_1 + n_2)(n_1 + n_2 - 2)} D^2 \frac{n_1 + n_2 - k - 1}{k} \qquad (2\text{-}6)$$

which is the same as (2-2). Thus the output from a discriminant analysis program will permit us to test the hypothesis of equality of means without difficulty.

For the technical college data of Chapter 1, the D^2 between engineers and builders was 1.16, and the sample sizes were $n_1 = 404$, $n_2 = 400$. Thus

$$F = \frac{404 \cdot 400}{804} \frac{804 - 3 - 1}{(802)(3)} \cdot 1.16$$

$$= 77.52$$

with 3 and 800 df. The .001 value of F with 3, 800 df $\approx$ 5.48 so the observed differences are highly significant.

Tests of sufficiency of a subset of variables

The second objective is to decide if all variables are needed. Stated another way, will a subset of k_1 of the k variables do as good a job as the whole set? Rao (*454*) gives methodology to make this decision. Suppose that a subset is specified and the variables $x_1 \cdots x_{k_1}$ are to be tested for sufficiency as discriminators.

Partition $\mathbf{x}$ into $(\mathbf{x}_1, \mathbf{x}_2)$, where $\mathbf{x}_1 = (x_1, \ldots, x_{k_1})$ and $\mathbf{x}_2 = (x_{k_1+1}, \ldots, x_k)$. Then Rao proves that the following statements are equivalent:

a. The random variable obtained by subtracting from $\mathbf{x}_2$ its regression on $\mathbf{x}_1$ has the same expected value in both the populations.

b. The coefficients of the components $x_{k_1+1}, \ldots, x_k$ in the linear discriminant function are all zero.

c. there is no additional distance contributed by the variables $x_{k_1+1}, \ldots, x_k$.

d. Every linear function of $\mathbf{x}$ uncorrelated with $\mathbf{x}_1$ has the same expected value for both the populations.

e. If $\mathbf{x} \sim N_k$ (that is, k variate normal), then the conditional distribution of $\mathbf{x}$ given $\mathbf{x}_1$ is the same for both the populations, which is the same as saying that $\mathbf{x}_1$ is sufficient for discrimination between the populations.

He derives the statistic

$$F = \frac{n_1 + n_2 - k - 1}{k - k_1} \frac{C(D_k^2 - D_{k_1}^2)}{1 + CD_{k_1}^2} \qquad (2\text{-}7)$$

where D_k^2 and $D_{k_1}^2$ are the Mahalanobis D^2 statistics on the full set and subset, respectively, and

$$C = \frac{n_1 n_2}{(n_1 + n_2)(n_1 + n_2 - 2)} \tag{2-8}$$

This has an F distribution with $k - k_1$ and $n_1 + n_2 - k - 1$ df. An important special case is $k_1 = k - 1$, for which we wish to determine if a single specified variable has no discriminating power. Then we can use

$$F = \frac{n_1 + n_2 - k - 1}{1} \frac{C(D_k^2 - D_{k-1}^2)}{1 + CD_{k-1}^2} \tag{2-9}$$

which has 1 and $n_1 + n_2 - k - 1$ df. This, of course, is t^2. If one wished to test all coefficients in this manner, one might find the t's and use Dunn's tables (*373a*).

This is only appropriate if one variable is to be eliminated. If a second variable is to be eliminated, the procedure should be repeated with the $k - 1$ remaining variables. This is a heuristic procedure and is similar to a stepdown procedure in regression. More will be said on variable selection in Chapter 6.

Rao has also given a test for a specified discriminant function (that is, $\mathbf{n}$ is specified by the user before seeing the data), $\mathbf{n}'\mathbf{x}$. The analogue to Mahalanobis distance is $D_A^2 = [\mathbf{n}'(\bar{\mathbf{x}}_1 - \bar{\mathbf{x}}_2)]^2/\mathbf{n}'\mathbf{S}\mathbf{n}$. Then the test statistic is

$$F = \frac{n_1 + n_2 - k - 1}{k - 1} C \left(\frac{D^2 - D_A^2}{1 + CD_A^2}\right) \tag{2-10}$$

where C is as before and F has $k - 1$ and $n_1 + n_2 - k - 1$ df. These results are conveniently given in Rao's 1970 article, although they antedate his 1952 book (*429, 436, 440*).

Let us refer to the technical college data to test for the sufficiency of x_1 and x_3. That is, we wish to test if the English score is superfluous. From Tables 1-1 through 1-3 we calculate

$$\begin{aligned} D_2^2 &= (7.23 \quad 2.09) \begin{pmatrix} 55.58 & 11.66 \\ 11.66 & 69.21 \end{pmatrix}^{-1} \begin{pmatrix} 7.23 \\ 2.09 \end{pmatrix} \\ &= (7.23 \quad 2.09) \begin{pmatrix} .01865 & -.00314 \\ -.00314 & .01498 \end{pmatrix} \begin{pmatrix} 7.23 \\ 2.09 \end{pmatrix} \\ &= .945 \end{aligned}$$

Next we calculate

$$C = \frac{404 \cdot 400}{804 \cdot 802} = .2506$$

and

$$F = \frac{800}{1} \frac{(.2506)(1.16 - .945)}{1 + (.2506)(.945)}$$

$$= 34.85$$

which, with 1 and 800 df, is highly significant. Thus, one would not wish to eliminate the English test in order to assign students to engineering or building trades. Similarly, one can find that if one does not use the arithmetic score, the D^2 becomes .500 and $F = 117.6$. If one does not use the form relations score, D^2 becomes 1.155 and $F = .78$. Thus the form relations score could be eliminated, but not the arithmetic or English scores.

Tests of discriminant coefficients that have a particular nonzero value are not useful in general, because the coefficients are determined only up to a constant multiple. That is, if $D_S(\mathbf{x})$ is a discriminant function, $mD_S(\mathbf{x})$, where m is a constant, is an equivalent one. The ratios of the coefficients are then unique. It is possible to give tests for various hypotheses about the ratios, as Rao does, but they are not frequently used in this day of stepwise analysis.

Estimation of error rates

The third objective in the evaluation of a discriminant function is to determine its performance in the classification of future observations. When the parameters are known, the error rate is

$$T(R,f) = p_1 \int_{R_2} f_1(\mathbf{x})\, d\mathbf{x} + p_2 \int_{R_1} f_2(\mathbf{x})\, d\mathbf{x} \qquad (2\text{-}11)$$

which is the optimum for these distributions. If $f_i(\mathbf{x})$ is multivariate normal with mean $\boldsymbol{\mu}_i$ and covariance $\boldsymbol{\Sigma}$, we can easily calculate these rates. Table 2-1 gives optimum error rates for some values of $p_1 = 1 - p_2$ and

Table 2-1 *Optimum error rates for normal distributions*

	p_1				
δ^2	*.1*	*.2*	*.3*	*.4*	*.5*
1	.098	.186	.253	.295	.309
4	.074	.112	.139	.154	.159
9	.034	.050	.060	.065	.067

$\delta^2 = (\boldsymbol{\mu}_1 - \boldsymbol{\mu}_2)'\boldsymbol{\Sigma}^{-1}(\boldsymbol{\mu}_1 - \boldsymbol{\mu}_2)$. Recall that P_1 and P_2 are given by (1-18). Error rates of this magnitude are the best that can be hoped for, given the values of δ^2 and p_i. Observe that the optimum error rate is never greater than p_1, for if it were, one could simply assign all observations to Π_2 and have a total error rate equal to p_1. This procedure is sometimes the optimal one for multinomial distributions.

What is to be done if the parameters are not known? A number of possible error rates may be defined, and each one, depending on the circumstances, may be valuable.

The function $T(R, f)$ defines the error rates. The first argument refers to the classification regions; the second argument is the presumed distribution of the observations that will be classified. The error rates of interest are as follows:

1. The optimum error rate:

$$T(R, f) \qquad \text{[defined by (2-11)]}$$

2. The error rate for the sample discriminant function as it will perform in future samples:

$$T(\hat{R}, f) = p_1 \int_{\hat{R}_2} f_1(\mathbf{x})\, d\mathbf{x} + p_2 \int_{\hat{R}_1} f_2(\mathbf{x})\, d\mathbf{x} \tag{2-12}$$

3. The expected error rate for discriminant functions based on samples of n_1 from Π_1 and n_2 from Π_2:

$$E(T(\hat{R}, f)) = E\left(p_1 \int_{\hat{R}_2} f_1(\mathbf{x})\, d\mathbf{x} + p_2 \int_{\hat{R}_1} f_2(\mathbf{x})\, d\mathbf{x}\right) \tag{2-13}$$

4. The plug-in estimate of the error rate obtained by using the estimated parameters for f_1 and f_2:

$$T(\hat{R}, \hat{f}) = p_1 \int_{\hat{R}_2} \hat{f}_1(\mathbf{x})\, d\mathbf{x} + p_2 \int_{\hat{R}_1} \hat{f}_2(\mathbf{x})\, d\mathbf{x} \tag{2-14}$$

5. The apparent error rate: This is defined as the fraction of observations in the initial sample which are misclassified by the sample discriminant.

Hills (*237*) calls the second rate the *actual error rate* and the third the *expected actual error rate*. Hills also shows that

$$E(T(\hat{R}, \hat{f})) < T(R, f) < T(\hat{R}, f) \tag{2-15}$$

Thus the actual error rate is greater than the optimum error rate, and, it, in turn, is greater than the expectation of the plug-in estimate of the error rate. Fukunaga and Kessel (*188*) prove a similar inequality.

To fix ideas, consider the multivariate normal distribution and assume that $p_1 = \frac{1}{2}$. The optimum classification region is

$$R_1 = \{\mathbf{x}: D_T(\mathbf{x}) = [\mathbf{x} - \tfrac{1}{2}(\boldsymbol{\mu}_1 + \boldsymbol{\mu}_2)]'\boldsymbol{\Sigma}^{-1}(\boldsymbol{\mu}_1 - \boldsymbol{\mu}_2) > 0\}$$

R_2 otherwise.

$$\begin{aligned} T(R, f) &= \frac{1}{2}\int_{R_2} f_1(\mathbf{x})\, d\mathbf{x} + \frac{1}{2}\int_{R_1} f_2(\mathbf{x})\, d\mathbf{x} \\ &= \tfrac{1}{2}\Pr\{D_T(\mathbf{x}) < 0 \mid \Pi_1\} + \tfrac{1}{2}\Pr\{D_T(\mathbf{x}) > 0 \mid \Pi_2\} \\ &= \Phi\left(-\frac{\delta}{2}\right) \end{aligned} \tag{2-16}$$

The actual error rate is

$$T(\hat{R}, f) = \frac{1}{2}\int_{\hat{R}_2} f_1(\mathbf{x})\, d\mathbf{x} + \frac{1}{2}\int_{\hat{R}_1} f_2(\mathbf{x})\, d\mathbf{x} \tag{2-17}$$

Now

$$\hat{R}_1 = \{\mathbf{x}: D_S(\mathbf{x}) = [\mathbf{x} - \tfrac{1}{2}(\bar{\mathbf{x}}_1 + \bar{\mathbf{x}}_2)]'\mathbf{S}^{-1}(\bar{\mathbf{x}}_1 - \bar{\mathbf{x}}_2) > 0\}$$

The probability that $\mathbf{x}$ falls in R_1 if $\mathbf{x}$ belongs to Π_2 is

$$\begin{aligned} \Pr(D_S(\mathbf{x}) > 0 \mid \Pi_2) &= \Pr\left[\frac{D_S(\mathbf{x}) - D_S(\boldsymbol{\mu}_2)}{\sqrt{V_D}} > \frac{-D_S(\boldsymbol{\mu}_2)}{\sqrt{V_D}}\right] \\ &= \Phi\left[\frac{D_S(\boldsymbol{\mu}_2)}{\sqrt{V_D}}\right] \end{aligned} \tag{2-18}$$

where

$$V_D = (\bar{\mathbf{x}}_1 - \bar{\mathbf{x}}_2)'\mathbf{S}^{-1}\boldsymbol{\Sigma}\mathbf{S}^{-1}(\bar{\mathbf{x}}_1 - \bar{\mathbf{x}}_2)$$

and

$$\Pr(D_S(\mathbf{x}) < 0 \mid \Pi_1) = \Phi\left[\frac{-D_S(\boldsymbol{\mu}_1)}{\sqrt{V_D}}\right] \tag{2-19}$$

Thus

$$T(\hat{R}, f) = \tfrac{1}{2}\Phi\left[\frac{-D_S(\boldsymbol{\mu}_1)}{\sqrt{V_D}}\right] + \tfrac{1}{2}\Phi\left[\frac{D_S(\boldsymbol{\mu}_2)}{\sqrt{V_D}}\right] \tag{2-20}$$

The reader will note that this formula will not be of much use unless estimates of $\boldsymbol{\mu}_1$, $\boldsymbol{\mu}_2$, and $\boldsymbol{\Sigma}$ are used. However, it is a handy one to recall if one is doing sampling experiments and needs the actual error rates.

The expected actual error rate is found by determining the expected value of (2-20) over all possible samples of size n_1 and n_2. To obtain the plug-in error rate, one estimates for $\boldsymbol{\mu}_1$, $\boldsymbol{\mu}_2$, and $\boldsymbol{\Sigma}$ by $\bar{\mathbf{x}}_1$, $\bar{\mathbf{x}}_2$, and $\mathbf{S}$, respectively, which gives

$$D_S(\hat{\boldsymbol{\mu}}_1) = D_S(\bar{\mathbf{x}}_1) = [\bar{\mathbf{x}}_1 - \tfrac{1}{2}(\bar{\mathbf{x}}_1 + \bar{\mathbf{x}}_2)]'\mathbf{S}^{-1}(\bar{\mathbf{x}}_1 - \bar{\mathbf{x}}_2) = \tfrac{1}{2}D^2 \tag{2-21}$$

$$D_S(\hat{\boldsymbol{\mu}}_2) = (\bar{\mathbf{x}}_2 - \tfrac{1}{2}[\bar{\mathbf{x}}_1 + \bar{\mathbf{x}}_2)]'\mathbf{S}^{-1}(\bar{\mathbf{x}}_1 - \bar{\mathbf{x}}_2) = -\tfrac{1}{2}D^2 \tag{2-22}$$

$$\hat{V}_D = (\bar{\mathbf{x}}_1 - \bar{\mathbf{x}}_2)'\mathbf{S}^{-1}\hat{\mathbf{\Sigma}}\mathbf{S}^{-1}(\bar{\mathbf{x}}_1 - \bar{\mathbf{x}}_2) = (\bar{\mathbf{x}}_1 - \bar{\mathbf{x}}_2)'\mathbf{S}^{-1}(\bar{\mathbf{x}}_1 - \bar{\mathbf{x}}_2) = D^2 \tag{2-23}$$

So

$$T(\hat{R}, \hat{f}) = \tfrac{1}{2}\Phi\left(-\frac{D}{2}\right) + \tfrac{1}{2}\Phi\left(-\frac{D}{2}\right) = \Phi\left(-\frac{D}{2}\right) \tag{2-24}$$

This is the maximum likelihood estimate of the error rate in this case. It is thus consistent and asymptotically efficient. It is not unbiased, however, and for small samples may be quite poor.

Estimation of error rates has received considerable attention in the literature. An extensive bibliography has been published by Toussaint (*533*). The sampling distribution of $D_S(\mathbf{x})$ has been studied, and asymptotic expansions of the distribution are available (*398, 23, 24*). However, it is not definitely known how good these expansions are for small samples, so other methods need to be used. Hills (*237*), Lachenbruch (*311*), Lachenbruch and Mickey (*315*), and Dunn (*156*), among others, have considered this problem.

Hills discusses estimates of the expected actual error rate. He notes that the maximum likelihood estimates may be used and points out that a conservative confidence region can be constructed by getting a $(1 - \alpha)$-level confidence region for the parameters of $f_1(\mathbf{x})$ and $f_2(\mathbf{x})$ and determining the maximum and minimum values of the error rate. This interval then has confidence at least $1 - \alpha$. This may not always be a simple task in multivariate problems. He then comments on the leaving-one-out method originally proposed by Lachenbruch (*311*). One estimates the discriminant rule omitting one sample observation and then uses the rule to classify the point left out. This is done for all observations and the number of misclassifications tallied. This gives an almost unbiased estimate of the expected actual error rate (see the Appendix to this chapter). Hills also discusses the multinomial distribution and provides some inequalities for error rates. The apparent error rate has the advantage of being easy to calculate and not requiring distributional assumptions. However, it is seriously biased and underestimates the expected actual error rate. It was for this reason that the leaving-one-out method was originally proposed. The apparent error rate was first proposed by Smith (*491*).

Lachenbruch and Mickey (*315*) compared a number of methods for estimating error rates, which include:

1. The apparent error rate.

2. The estimated actual error rate [Eq. (2-24)].

3. Using other estimates of δ^2 in Eq. (2-24).

4. The leaving-one-out method.

5. Using estimates of δ^2 in Okamoto's expansion of the distribution of $D_S(\mathbf{x})$.

Another suggested method is to partition the sample into two parts: use one part to construct the rule and the other to evaluate it. This method is wasteful of data, needs large samples that are often not readily available, and does not evaluate the discriminant function that will be used in practice. For these reasons this method was not evaluated by Lachenbruch and Mickey. Sampling experiments were used to obtain some idea of the performance of these rules under a wide set of parameter combinations assuming normality. It was found that using D_S^2 to estimate δ^2 in Okamoto's expansions did better than any other method, where

$$D_S^2 = \frac{n_1 + n_2 - k - 3}{n_1 + n_2 - 2} D^2$$

The second best estimate was to use D^2 for δ^2 in the expansion. Then came using $\Phi(-D_S/2)$ followed by the leaving-one-out method. The apparent error rate and the estimated actual error rate (the MLE) were the poorest of the lot, being highly biased. In further studies it was indicated that if the assumption of normality does not hold, the methods using that assumption must be discarded. Then only the apparent error rate, the leaving-one-out method, and the sample partition methods remain. For moderately large samples, the apparent error rate may be used, but for small samples, the leaving-one-out method seems preferable.

It might be noted that no method of those suggested estimates the actual error rate. Most of them estimate the expected actual error rate. The method of plugging in the maximum likelihood estimate of the parameters underestimates the optimum error rate on the average, and while consistent, can always be improved upon (in the normal case at least).

Lachenbruch (*314*) suggested another estimate for the expected actual error rate based on the expected mean and variance of $D_S(\mathbf{x})$. The expected mean is

$$E(D_S(\boldsymbol{\mu}_i)) = \tfrac{1}{2}C_1 \left[\delta^2(-1)^{i+1} - \frac{k(n_2 - n_1)}{n_1 n_2}\right] \qquad (2\text{-}25a)$$

where

$$C_1 = \frac{n_1 + n_2 - 2}{n_1 + n_2 - k - 3}$$

and

$$E(V_D) = C_2 \left[\delta^2 + \frac{k(n_1 + n_2)}{n_1 n_2} \right] \tag{2-25b}$$

where

$$C_2 = \frac{(n_1 + n_2 - 3)(n_1 + n_2 - 2)^2}{(n_1 + n_2 - k - 2)(n_1 + n_2 - k - 3)(n_1 + n_2 - k - 5)}$$

Thus one may get approximate values of P_1 and P_2 as

$$\bar{P}_1 = \Phi \left[\frac{\ln \dfrac{1 - p_1}{p_1} - E(D_s(\boldsymbol{\mu}_1))}{\sqrt{E(V_D)}} \right]$$

and

$$\bar{P}_2 = \Phi \left[- \frac{\ln \dfrac{1 - p_1}{p_1} + E(D_S(\boldsymbol{\mu}_2))}{\sqrt{E(V_D)}} \right] \tag{2-26}$$

These may be estimated by estimating δ^2 by D^2 in (2-25a) and (2-25b). The behavior of this estimate is comparable to using Okamoto's expansion with D^2 estimating δ^2. This approach was used to determine the necessary sample size for a discriminant function to be within a given tolerance of its optimum error rate. Some results from this study were given in Table 1-9.

Sorum (*500*) has considered estimating the actual probability of misclassification in the normal case, which she refers to as the conditional probability of misclassification. She also studied the situation when additional m_2 observations from Π_2 were obtained and found that for this case the conditional uniform minimum variance unbiased estimator of P_2 was

$$\hat{P}_2 = 1 - \Phi \left\{ \left(1 - \frac{1}{m_2}\right) \left[\tfrac{1}{2} D + \frac{1}{D} (\bar{\mathbf{x}}_1 - \bar{\mathbf{x}}_2)' \boldsymbol{\Sigma}^{-1} (\bar{\mathbf{x}}_2 - \bar{\mathbf{t}}_2) \right] \right\} \tag{2-27}$$

where $\bar{\mathbf{t}}_2$ is the mean of the additional m_2 observations. Note that she assumed that $\boldsymbol{\Sigma}$ was known. She observed that plausible estimates could be obtained for all estimates given by Lachenbruch and Mickey (*315*) by using $\bar{\mathbf{t}}_2$ to estimate $\boldsymbol{\mu}_2$. Similar results hold, of course, for P_1. The properties of the estimates indicate that use of the normality assumption when justified leads to a better estimate in the sense that the asymptotic mean-square error is less than that for those in which normality did not hold.

Dunn (*156*) did a Monte Carlo study of the unconditional probability of correct classification. In the terminology adopted here, this is one minus the expected actual error rate. She gives the estimates of this quantity for a variety of parameter combinations. She also compares the estimates given in (2-26) with her results and notes that they tend to be slightly conservative. In no case were they more than .02 less than the Monte Carlo estimates for the probability of correct classification.

Fukunaga and Kessel (*189*) discuss the use of the apparent error rate and the leaving-one-out method (referred to as the C method and the L method, respectively) and indicate how the leaving-one-out method may be used for quadratic discriminants. Using the Parzen approach (*406*), they also extend the two methods to nonparametric density estimation and give some experimental results. In a set of data with $k = 8$, $n_1 = n_2$, and optimal error rate .019, they found errors as in Table 2-2. Their results indicate that for large samples the methods are about the same, although the bias in the apparent error rate is still obvious. The interesting feature here is the atrocious performance of the apparent error rate when applied in the nonparametric situation. The data used were for a quadratic discriminant analysis (that is, were normal with difference covariance matrices).

Returning to the technical college data, we find very little difference in the methods of error rate estimation that could be used. The simplest method gave $\Phi(-D/2) = .2951$; we also computed $\Phi(-D_s/2) = .2956$, and from (2-26) (assuming that $p_1 = \frac{1}{2}$), $\bar{P}_1 = .2968$. Thus these estimates are within .002 of one another, a trivial difference. When large samples are available, it makes no sense to try the fancier estimates, as the bias has already been reduced to a very low level.

The major portion of this chapter has been concerned with methods based on the normal distribution. If the data are normal, or nearly so, these methods work satisfactorily. Difficult theoretical problems arise when the

Table 2-2 *Error-rate estimation for discrimination example*[a]

	Quadratic discriminant		*Nonparametric density estimate*	
$n_1 = n_2$	*Apparent*	*Leaving-one-out*	*Apparent*	*Leaving-one-out*
100	.0144	.0215	.001	.029
200	.0156	.0200	.0045	.0235
400	.0183	.0197	.0065	.022

Source: Fukunaga and Kessel (*189*).
[a] Optimum error rate, .019.

observations are not normally distributed. First, the investigator usually has little or no idea of the form of the distribution. Second, finding the distribution of linear combinations of nonnormal variables is a difficult and as yet unsolved problem. None of the error-estimation methods based on normality can be expected to work. However, the apparent error rate and leaving-one-out method seem to be adequate, the former only if large samples are available.

Appendix

It is useful to have a simple expression for the leaving-one-out method in a discriminant analysis program. Note that for either the linear or quadratic functions we can write the discriminant function as

$$\tfrac{1}{2}[Q_2(\mathbf{x}) - Q_1(\mathbf{x})] + \tfrac{1}{2}\ln\frac{|\mathbf{\Sigma}_2|}{|\mathbf{\Sigma}_1|} \tag{2-28}$$

where $Q_i(\mathbf{x}) = (\mathbf{x} - \boldsymbol{\mu}_i)'\mathbf{\Sigma}_i^{-1}(\mathbf{x} - \boldsymbol{\mu}_i)$. In the equal covariance case, the term involving the logarithm is zero, and the quadratic terms in $\mathbf{x}$ vanish. If $\mathbf{\Sigma}_1 \neq \mathbf{\Sigma}_2$, and we estimate $\boldsymbol{\mu}_i$ by $\bar{\mathbf{x}}_i$ and $\mathbf{\Sigma}_i$ by $\mathbf{S}_i$, the within-groups mean and covariance matrices, we have the sample quadratic discriminant function.

To obtain the apparent error rate we calculate

$$\begin{aligned} Q_{11}(\mathbf{x}_i) &= (\mathbf{x}_i - \bar{\mathbf{x}}_1)'\mathbf{S}_1^{-1}(\mathbf{x}_i - \bar{\mathbf{x}}_1) \\ Q_{22}(\mathbf{x}_i) &= (\mathbf{x}_i - \bar{\mathbf{x}}_2)'\mathbf{S}_2^{-1}(\mathbf{x}_i - \bar{\mathbf{x}}_2) \end{aligned} \tag{2-29}$$

and

$$Q_A(\mathbf{x}_i) = \tfrac{1}{2}[Q_{22}(\mathbf{x}_i) - Q_{11}(\mathbf{x}_i)] + \tfrac{1}{2}\ln\frac{|\mathbf{S}_2|}{|\mathbf{S}_1|} \tag{2-30}$$

for each i and count the number of observations misclassified. For the leaving-one-out method we calculate

$$\begin{aligned} Q_L(\mathbf{x}_i) = Q_A(\mathbf{x}_i) - \frac{1}{2}\left\{\frac{Q_{11}(\mathbf{x}_i) + Q_{11}^2(\mathbf{x}_i)}{n_1 - 1 - Q_{11}(\mathbf{x}_i)}\right. \\ \left. + \ln\left[1 - \frac{1}{n_1 - 1}Q_{11}(\mathbf{x}_i)\right] + k\ln\frac{n_1}{n_1 - 1}\right\} \end{aligned} \tag{2-31}$$

if $\mathbf{x}_i$ comes from Π_1 and

$$\begin{aligned} Q_L(\mathbf{x}_i) = Q_A(\mathbf{x}_i) + \frac{1}{2}\left\{\frac{Q_{22}(\mathbf{x}_i) + Q_{22}^2(\mathbf{x}_i)}{n_2 - 1 - Q_{22}(\mathbf{x}_i)}\right. \\ \left. + \ln\left[1 - \frac{1}{n_2 - 1}Q_{22}(\mathbf{x}_i)\right] + k\ln\frac{n_2}{n_2 - 1}\right\} \end{aligned} \tag{2-32}$$

if $\mathbf{x}_i$ comes from Π_2. This is done for all $\mathbf{x}_i$ and the number of misclassifications is determined.

For the linear discriminant function we need

$$
\begin{aligned}
Q_{12}(\mathbf{x}_i) &= (\mathbf{x}_i - \bar{\mathbf{x}}_1)'\mathbf{S}^{-1}(\mathbf{x}_i - \bar{\mathbf{x}}_2) \\
Q_{11}(\mathbf{x}_i) &= (\mathbf{x}_i - \bar{\mathbf{x}}_1)'\mathbf{S}^{-1}(\mathbf{x}_i - \bar{\mathbf{x}}_1) \\
Q_{22}(\mathbf{x}_i) &= (\mathbf{x}_i - \bar{\mathbf{x}}_2)'\mathbf{S}^{-1}(\mathbf{x}_i - \bar{\mathbf{x}}_2)
\end{aligned}
\tag{2-33}
$$

where $\mathbf{S}$ is the pooled covariance matrix. The apparent error rate can be found by calculating

$$Q_A(\mathbf{x}_i) = \tfrac{1}{2}[Q_{22}(\mathbf{x}_i) - Q_{11}(\mathbf{x}_i)] \tag{2-34}$$

This is not the most efficient way of doing it; it is done this way to illustrate the leaving-one-out method, which calculates

$$
\begin{aligned}
Q_L(\mathbf{x}_i) = \frac{1}{2}\Bigg\{&\frac{\nu - 1}{\nu} Q_{22}(\mathbf{x}_i) + \frac{C_1(\nu - 1)}{\nu^2} \frac{[Q_{12}(\mathbf{x}_i)]^2}{1 - (C_1/\nu)Q_{11}(\mathbf{x}_i)} \\
&- C_1{}^2\left[\frac{\nu - 1}{\nu} Q_{11}(\mathbf{x}_i) + \frac{C_1(\nu - 1)}{\nu^2} \frac{Q_{11}^2(\mathbf{x}_i)}{1 - (C_1/\nu)Q_{11}(\mathbf{x}_i)}\right]\Bigg\}
\end{aligned}
\tag{2-35}
$$

if $\mathbf{x}_i$ comes from Π_1 and

$$
\begin{aligned}
Q_L(\mathbf{x}_i) = \frac{1}{2}\Bigg\{&C_2{}^2\left[\frac{\nu - 1}{\nu} Q_{22}(\mathbf{x}_i) + C_2 \frac{\nu - 1}{\nu} \frac{Q_{22}(\mathbf{x}_i)}{1 - (C_2/\nu)Q_{22}(\mathbf{x}_i)}\right. \\
&\left. - \frac{\nu - 1}{\nu} Q_{11}(\mathbf{x}_i) - \frac{C_2(\nu - 1)}{\nu^2} \frac{Q_{12}^2(\mathbf{x}_i)}{1 - (C_2/\nu)Q_{22}(\mathbf{x}_i)}\right]\Bigg\}
\end{aligned}
\tag{2-36}
$$

if x_i comes from Π_2. Here $C_i = n_i/n_i - 1$ and $\nu = n_1 + n_2 - 2$.

The results for the quadratic discriminant function were originally given by Fukunaga and Kessel (*188*); the results for the linear discriminant function were first suggested by Lachenbruch in a different form.

Problems

1. Use the data of Table 1-3 to test the hypothesis that the means of each pair of trades are identical.
2. Verify that the F tests for the superfluousness of the arithmetic test and the form relations tests are 117.6 and 0.78.
3. Although tests for specific values of discriminant functions are not generally useful, if one specifies that the coefficients will be $\mathbf{b} = \mathbf{S}^{-1}(\bar{\mathbf{x}}_1 - \bar{\mathbf{x}}_2)$, an approximate standard error can be obtained. If we

assume that the sample sizes are large enough that $\mathbf{S} \approx \mathbf{\Sigma}$, then the covariance matrix of $\mathbf{b}$ is $(1/n_1 + 1/n_2)\mathbf{S}^{-1}$. Show this.

4. Verify the entries in Table 2-1 for
   (a) $p_1 = .2, \quad \delta^2 = 4$
   (b) $p_1 = .3, \quad \delta^2 = 1$
   (c) $p_1 = .4, \quad \delta^2 = 9$
5. For the engineers/builders data of Chapter 1, assume that $p_1 = \frac{1}{2}$,

$$\boldsymbol{\mu}_1' = (28, \quad 98, \quad 34)$$

$$\boldsymbol{\mu}_2' = (21, \quad 85, \quad 32)$$

$$\mathbf{\Sigma} = \begin{pmatrix} 50 & 25 & 0 \\ 25 & 300 & 0 \\ 0 & 0 & 60 \end{pmatrix}$$

   (a) Find $T(\hat{R}, f)$, the actual error rate, for the sample discriminant function computed in Chapter 1.
   (b) Find $\bar{P}_1$ and $\bar{P}_2$ by use of (2-26).
6. (a) Show that if $\mathbf{B} = \mathbf{A} + \mathbf{uv}'$, where $\mathbf{A}$ is nonsingular and $\mathbf{u}$ and $\mathbf{v}$ are column vectors, then

$$\mathbf{B}^{-1} = \mathbf{A}^{-1} - \frac{\mathbf{A}^{-1}\mathbf{uv}'\mathbf{A}^{-1}}{1 + \mathbf{v}'\mathbf{A}^{-1}\mathbf{u}}$$

   (b) Let $\bar{\mathbf{x}}_{1(j)}$ be the mean of the $n_1 - 1$ observations from Π_1, omitting the jth observation. Show that

$$\bar{\mathbf{x}}_{1(j)} = \bar{\mathbf{x}}_1 - \frac{\mathbf{x}_{1j} - \bar{\mathbf{x}}_1}{n_1 - 1} = \frac{\bar{\mathbf{x}}_1 - \mathbf{u}_j}{n_1 - 1}$$

   (c) Show that

$$(n_1 + n_2 - 3)\mathbf{S}_{(j)} = (n_1 + n_2 - 2)\mathbf{S} - \frac{n_1}{n_1 - 1}(\mathbf{u}_j\mathbf{u}_j')$$

   (d) Use (a), (b), and (c) to show that the leaving-one-out method computes the following discriminant function when x comes from Π_1:

$$D_{(1j)}(\mathbf{x}_{1j}) = \frac{n_1 + n_2 - 3}{n_1 + n_2 - 2}\left\{\mathbf{x}_{1j} - \tfrac{1}{2}(\bar{\mathbf{x}}_1 + \bar{\mathbf{x}}_2) + \frac{\mathbf{u}_j}{2(n_1 - 1)}\right\} \mathbf{S}_{(j)}^{-1}(\bar{\mathbf{x}}_1 - \bar{\mathbf{x}}_2) - \frac{\mathbf{u}_j}{n_1 - 1}$$

Projects

1. Modify the program that you wrote for Chapter 1, project 2, to obtain the apparent error rate to calculate $\Phi(-D_S/2)$, $\bar{P}_1$, P_2 from (2-26) and the leaving-one-out error rate. Calculate the test given by (2-9) for each variable. Print the approximate standard errors of the discriminant coefficients, $[s^{ii}(1/n_1 + 1/n_2)]^{1/2}$.
2. Modify the program that you wrote for Chapter 1, project 3, to obtain the apparent error rate and the leaving-one-out error rate. Can you think of other methods to use for estimating error rates in the quadratic case?

3 Robustness of the linear discriminant function

The linear discriminant function is the assignment rule when the following assumptions are satisfied:

1. $f_1(\mathbf{x})$ and $f_2(\mathbf{x})$ are multivariate normal.

2. The covariance matrix in Π_1 is the same as the covariance matrix in Π_2.

3. The a priori probabilities, p_1 and p_2, are known.

4. The means, $\boldsymbol{\mu}_1$ and $\boldsymbol{\mu}_2$, and covariance matrix, $\boldsymbol{\Sigma}$, are known.

If one or more of these conditions do not hold, the discriminant function we calculate will not be the optimum assignment rule. If the means and covariance matrix are unknown, we must estimate them by using a sample. In this case, two additional problems may arise. The initial samples may not be correctly assigned, and there may be missing values. This chapter is concerned with the behavior of the classification rule obtained from the linear discriminant function when one of the assumptions is violated.

In Chapter 2 we discussed the effects on the error rate when estimates are substituted for the unknown parameters. In general, if the p_i are unknown, one may estimate them from the sample if the sample is chosen from a whole population rather than two separate samples taken from two subgroups. Otherwise, one may use a minimax approach to the problem.

If the distribution of $\mathbf{x}$ is not multivariate normal, the optimal assignment rule is, in general, not the linear discriminant function. The results given in this chapter will deal with the use of the linear discriminant function for a set of Bernoulli variables, for bivariate Poisson and negative multinomial variables, and for some continuous nonnormal distributions. Violation of the equal covariance assumption was studied in an article by Gilbert (*204*).

When using a sample discriminant function, $D_S(\mathbf{x})$, we make an important assumption that the initial samples were originally assigned correctly. A section of this chapter deals with studies of violations of this assumption. Finally, some studies of the treatment of missing values are discussed.

Nonnormal data

A number of studies have been concerned with the behavior of various methods of discrimination when qualitative variables are used. In many fields the data are observations of the presence or absence of an attribute. In medicine, for example, this may have the form of a checklist of symptoms or historical data. The vector of responses clearly does not have a multivariate normal distribution. Indeed, the set of possible responses has a multinomial distribution on 2^k cells, where k is the number of Bernoulli (that is, $x_j = 0$ or 1) responses. By using a linear discriminant on the k Bernoulli variables, a considerable simplification may be obtained. But since the function obtained is not optimum, it is worthwhile to inquire how good it is. Various other techniques have been proposed and evaluated for use with Bernoulli variables.

Gilbert (*203*) studied the problem of classification for a variety of possible multivariate Bernoulli distributions. A multivariate Bernoulli distribution is the distribution of a set of Bernoulli variables. She considered five classification procedures:

1. The use of the estimated ratio of the multinomial probabilities n_{2i}/n_{1i}, where n_{ji} is the number of members of Π_j in multinomial cell i, $i = 1, \ldots, 2^k$. This is appropriate only if $n_{1+} = \Sigma n_{1i} = \Sigma n_{2i} = n_{2+}$. If not, then $(n_{2i}/n_{2+})/(n_{1i}/n_{1+})$ should be used. This is asymptotically optimal, but involves the estimation of $2^k - 1$ probabilities. If k is at all large, very large samples are needed (e.g., if $k = 10$, there are 1023 probabilities to estimate).

2. Assume that

$$\ln \frac{p_{\mathbf{x}2}}{p_{\mathbf{x}1}} = 2\boldsymbol{\beta}'\tilde{\mathbf{x}} \tag{3-1}$$

where $p_{\mathbf{x}j}$ is the probability of obtaining $\mathbf{x}$ when $\mathbf{x}$ belongs to Π_j and $\tilde{\mathbf{x}} = (1, \mathbf{x})$. Estimate $\boldsymbol{\beta}$ by maximum likelihood.

3. Under the same model as in procedure 2, obtain the minimum logit χ^2 estimates of $\boldsymbol{\beta}$, which minimize

$$\Sigma \frac{n_{\mathbf{x}2}n_{\mathbf{x}1}}{n_{\mathbf{x}}} \left(\ln \frac{n_{\mathbf{x}2}}{n_{\mathbf{x}1}} - 2\boldsymbol{\beta}'\tilde{\mathbf{x}}\right)^2 \tag{3-2}$$

4. Assume that the Bernoulli variables are independent, and find maximum likelihood estimates of $p_{\mathbf{x}2}/p_{\mathbf{x}1}$.

5. Use the linear discriminant function to assign to Π_1 or Π_2.

The first method is completely general but breaks down in practice because of the many parameters that must be estimated. The fourth procedure involves the generally untenable assumption of mutual independence, but as it has often been used in practice, it was evaluated. The second and third methods are intermediate to methods 1 and 4; that is, they allow some kinds of dependence, but not the completely general full multinomial model. If we denote

$$q_x = \Pr(\Pi_2 \mid \mathbf{x}) = \frac{p_{\mathbf{x}2}}{p_{\mathbf{x}2} + p_{\mathbf{x}1}}$$

then

$$\ln \frac{p_{\mathbf{x}2}}{p_{\mathbf{x}1}} = \ln \frac{q_{\mathbf{x}}}{1 - q_{\mathbf{x}}} = \text{logit } q_{\mathbf{x}} \tag{3-3}$$

This is assumed to be a linear function of $\mathbf{x}$ in models 2 and 3. If one defines $X_{k+1} = 0$ in Π_1 and $X_{k+1} = 1$ in Π_2, Gilbert noted that $\ln p\tilde{\mathbf{x}}$ could be written as

$$\ln p_{\tilde{\mathbf{x}}} = \alpha + \sum_{j=1}^{k+1} (-1)^{x_j} \alpha_k + \sum_{j<l} (-1)^{x_j + x_l} \alpha_{jl} + \cdots$$
$$+ (-1)^{\Sigma x_j} \alpha_{12 \cdots k(k+1)} \tag{3-4}$$

For the experiments she performed, she assumed that all interactions of order 2 or higher were zero, so

$$\ln p_{\tilde{\mathbf{x}}} = \alpha + \sum_{j=1}^{k+1} (-1)^{x_j} \alpha_j + \sum_{j<l} (-1)^{x_j + x_l} \alpha_{jl} \tag{3-5}$$

In this case it can be shown that (3-1) is satisfied and thus methods 2 and 3 are asymptotically optimum.

In addition to use of the probability of misclassification, she proposed use of the correlation between the optimum procedure and the sample procedure as a criterion for evaluating the method. She did this because the probability of misclassification considers only the case in which costs are assumed to be equal. She first compared the discriminant function with the optimum procedure based on the first-order interaction models of (3-5). Table 3-1 gives the distribution of the correlation coefficients thus found. A wide variety of populations was chosen within the framework of the model in (3-5). For $k = 2$, 540 cases were chosen; $k = 3$, 4596 cases; $k = 6$, 6976 cases.

Sampling experiments were performed on 15 populations using $n = 100$ or $n = 500$ and $k = 6$. In this study n_i, the number from Π_i was also a random variable, and it is not possible to determine what the a priori

Table 3-1 Frequency distribution of the sample correlation between optimum rule and linear discriminant function

ρ	$k = 2$	$k = 3$	$k = 6$
.99–1.0	.926	.795	.586
.98–.99	.059	.103	.168
.97–.98	.015	.035	.074
.96–.97	0	.020	.054
.95–.96	0	.014	.025
<.95	0	.033	.093

Source: Gilbert (*203*).

probabilities were in these data. Of course, $n_1 + n_2 = n$. The correlation and probability of misclassification were computed for each population and for each method.

Gilbert concluded that if the parameters of the distribution are "moderate," the LDF and the optimum procedure are very highly correlated. Based on the correlations and on the probability of misclassification, methods 2 to 5 are superior to method 1, the multinomial model.

A similar study has been performed by Moore (*376*) using a different interaction model, one based on the Bahadur model for Bernoulli variables. Bahadur (*32*) showed that the multinomial probabilities of the 2^k cells can be expressed in the following manner. Let

$$p_{ij} = E_i(x_j) \text{ the expectation of the } j\text{th variable when } \mathbf{x} \text{ comes from } \Pi_i$$

$$Z_{ij} = \frac{x_j - p_{ij}}{\sqrt{p_{ij}(1 - p_{ij})}}$$

$$r_i(jk) = E(Z_{ij}Z_{ik}) \tag{3-6}$$

$$r_i(jkl) = E(Z_{ij}Z_{ik}Z_{il}), \text{ etc.}$$

Then

$$\Pr(\mathbf{x} \mid \Pi_i) = \prod_{j=1}^{k} p_{ij}(1 - p_{ij})^{1-x_j}\Big[1 + \sum_{j<k} r_i(jk)Z_{ij}Z_{ik} + \sum_{j<k<l} r_i(jkl)Z_{ij}Z_{ik}Z_{il} + \cdots + r_i(1, 2, \ldots, k)Z_{i1}\cdots Z_{ik}\Big] \tag{3-7}$$

This provides a way of expressing a model in terms of means and correlations. A "first-order" model omits all terms involving correlations; a "second-order" model retains only those correlation terms of the form $r_i(jk)$.

Moore's study also evaluated five procedures:

1. The full multinomial model.

2. The procedure based on the first-order model (that is, independent variables).

3. The procedure based on the second-order model.

4. The linear discriminant function.

5. The quadratic discriminant function.

Procedures 1, 2, and 4 were also studied by Gilbert. As Gilbert did, Moore considered both the correlation of the sample procedure with the optimum and the probability of misclassification as criteria for evaluation.

Moore's study used $k = 6$ and sampled 19 pairs of populations. He sampled 50 replicates with $n_1 = n_2 = 50$ and 50 replicates with $n_1 = n_2 = 100$. Note that some cell must have zero frequency when $n_1 = n_2 = 50$. His conclusions generally agreed with Gilbert's. He found that for populations with zero correlation the first order and discriminant function methods did the best. If only a single pair of variables had a nonzero correlation, again the first-order and discriminant function procedures were best. For populations in which all correlations were positive, the discriminant model did somewhat more poorly than the full multinomial model. In the worst case the optimum error rate was .179, and the discriminant function was 11.1 percent poorer; its mean error rate was .199. This behavior is explained by reversals in the log-likelihood ratio; i.e., the log-likelihood ratio is not a monotonic function of the sum of x_j. He also provided some data taken from a multiphasic screening study in which the linear discriminant function and the first-order model performed best.

Moore cautioned that one must be aware of possible reversals in the log-likelihood ratio. The second-order procedure had limited usefulness, as did the quadratic procedure.

Martin and Bradley (*348*) suggested an alternative model for the multivariate binomial case which uses orthogonal polynomials. However, they did not compare the performance of this method to the methods that seem to be good, namely, the independent variables or linear discriminant function models.

Cochran and Hopkins (*115*) studied the classification of multivariate qualitative data but were not concerned with the linear discriminant function. Their work will be discussed in Chapter 4. Brown (*71*) performed a study similar to Gilbert's and Moore's which also confirmed the value of the independent variables and linear discriminant function methods.

Revo (*460*) studied various classification rules for some univariate and bivariate discrete distributions. The distributions were the Poisson, negative multinomial, and discretized normal. The discretized normal was constructed by partitioning each variable of a univariate or bivariate normal distribution into six categories. The goal of the study was to compare methods when the categories were ordered. The procedures used were:

1. The linear discriminant function.

2. The multinomial procedure. For the Poisson and negative multinomial distributions, the multinomial model was truncated at a point at which the probability of exceeding the truncation state was effectively zero.

3. The nearest-neighbor procedure (see Chapter 4).

4. The likelihood ratio procedure based on the model [that is, estimate the parameters of f_1 and f_2 and assign on the basis of $\hat{f}_1(\mathbf{x})/\hat{f}_2(\mathbf{x})$].

Sampling experiments were performed on several parameter combinations for each distribution. The probability of misclassification was used as the criterion for evaluation of the procedures, and the theoretical, actual, expected actual, and apparent error rates were calculated. Samples of size $n_1 = n_2 = 10$, 20, or 50 were used for the univariate experiments, and $n_1 = n_2 = 20$ or 50 in the bivariate experiments. It was found that the discriminant function compared favorably to the optimal rule in the great majority of the sampling experiments performed. The nearest-neighbor and multinomial rules were poorer, particularly for smaller sample sizes, because of the large sampling error involved with estimating many parameters. The likelihood ratio procedure based on the model performed in general as well as the discriminant function. Greater correlation between variates decreased the discriminating power of the second variate for the discriminant function method but increased it for the multinomial and nearest-neighbor procedures. If very large samples are available, the multinomial rule is satisfactory.

The general indications seem to be that the linear discriminant function performs fairly well on discrete data of various types. This may be because the log-likelihood ratio tends to be a monotonic function of the sum of the observations. There is also the possibility that it is robust, because one may, in choosing a cutoff point, vary the point widely without changing the error rate.

Relatively little has been done regarding the robustness of the discriminant function to continuous nonnormal distributions. Lachenbruch, Sneeringer, and Revo (*316*) considered a very simple version of the problem. It was assumed that all variables x_i were transformed normal variables y_i such that

$$
\begin{aligned}
y_i &= \log x_i, & 0 < x_i < \infty & \qquad \text{log normal} \\
y_i &= \log \frac{x_i}{1 - x_i}, & 0 < x_i < 1 & \qquad \text{logit normal} \qquad (3\text{-}8)
\end{aligned}
$$

or

$$y_i = \sinh^{-1}(x_i), \quad \infty < x_i < \infty \quad \sinh^{-1} \text{ normal}$$

It was also assumed that the covariance matrix of **y** corresponding to the **x** variable was the identity matrix but that the two populations had different means. Because of the nonlinear transformations, the **x**'s do not have the same covariance matrices. The authors found that for these distributions, the total error rate was often greatly increased, and the individual error rates were distorted in such a way that one was increased substantially above the optimum rate and the other decreased similarly. The effect was smallest for the logit-normal transformed group, possibly because it is bounded. To estimate error rates, with the sample sizes used here ($n_1 = n_2 = 100$ or $n_1 = 150$, $n_2 = 50$), the apparent error rate was satisfactory, although slightly biased. The leaving-one-out method did quite well, and performed slightly better than the apparent error rate. All methods based on the normal were very poor. Because of the distortion in error rates, approximate minimax rules were explored and found to reduce the distortion considerably. The behavior of quadratic discriminants was very poor. Using the apparent error rate for these was particularly bad. These results stand in sharp contrast to those for discrete distributions, which indicated fairly robust behavior of the linear discriminant function.

Unequal covariance matrices

The assumption of equal covariances may also be violated. If the covariance matrices are not equal, then the optimal rule is a quadratic discriminant function. Gilbert (*204*) studied the particular case in which one covariance matrix is a multiple of the other. She assumed that $\mathbf{\Sigma}_1 = \mathbf{I}$ and $\mathbf{\Sigma}_2 = d\mathbf{I}$. In this case, the linear discriminant function becomes

$$\begin{aligned} D(x) &= \mathbf{X}'[p_1 + (1 - p_1)d]^{-1}(\boldsymbol{\mu}_1 - \boldsymbol{\mu}_2) \\ &\quad - \tfrac{1}{2}(\boldsymbol{\mu}_1 - \boldsymbol{\mu}_2)'(\boldsymbol{\mu}_1 - \boldsymbol{\mu}_2)[p_1 + (1 - p_1)d]^{-1} \end{aligned} \qquad (3\text{-}9)$$

For various values of d, p_1, and δ^2 Gilbert compared the error rates when using the quadratic instead of the linear discriminant function. Table 3-2 gives some of her results. For large values of δ^2 the quadratic and linear functions behave similarly. If d is far from 1.0, the linear discriminant is poorer than the quadratic. This study assumed that the parameters were known, so these are results about the optimum error rate.

Marks and Dunn (*345*), assuming unknown parameters, compared the sample linear discriminant function, the best linear classification rule when unequal covariance matrices are present and the sample quadratic discriminant function with respect to their probabilities of misclassification. The best linear function for unequal covariance discrimination was proposed

Table 3-2 Improvement in optimum error rates with quadratic instead of linear discriminant functions

		k = 6				k = 10			
p_1	δ^2	d = .2	.5	2	5	d = .2	.5	2	5
.5	0	.41	.22	.22	.41	.46	.26	.26	.46
	1	.18	.07	.07	.18	.23	.12	.12	.23
	4	.09	.03	.03	.09	.11	.05	.05	.11
.667	0	.24	.06	.10	.25	.29	.11	.14	.29
	1	.20	.08	.05	.14	.25	.12	.08	.19
	4	.08	.03	.03	.07	.10	.05	.04	.10

Source: Gilbert (*204*).

by Anderson and Bahadur (*25*) and by Clunies-Ross and Riffenburgh (*107*). This rule assigns $\mathbf{x}$ to Π_2 if

$$\mathbf{x}'\mathbf{b} > \mathbf{b}'\bar{\mathbf{x}}_1 + t_1\mathbf{b}'\mathbf{S}_1\mathbf{b} \qquad (3\text{-}10)$$

where $\mathbf{b} = (t_1\mathbf{S}_1 + t_2\mathbf{S}_2)^{-1}(\bar{\mathbf{x}}_2 - \bar{\mathbf{x}}_1)$ and t_1 and t_2 are chosen to minimize the error rate.

Marks and Dunn used a slightly wider variety of situations than Gilbert did, but included some of her parameter combinations. The sampling results indicate that the linear function is quite satisfactory if the covariance matrices are not too different. In particular, the quadratic discriminant function is very poor for small sample sizes. When differences between covariance matrices are quite large, and the samples are also quite large, they recommend the use of the quadratic function.

Initial misclassification

The assumption that the initial samples from Π_1 and Π_2 are correctly classified may not always hold. Clerical errors may have occurred, or knowledge of the field may be such that there is an area of inaccuracy in initial assignment. For example, in studying adverse drug reactions, an investigation may find it quite difficult to decide if a mild complaint such as a headache is due to the drug administered or to some other factor. Lachenbruch (*312*) studied the effects on the error rate of the linear discriminant function if a fraction α_1 of the n_1 observations purportedly from Π_1 were really from Π_2, and a fraction α_2 of the n_2 observations were really from Π_1. Then for large samples the discriminant function tends to

$$D_M(\mathbf{x}) = K\left[D_T(\mathbf{x}) + \frac{\alpha_1 - \alpha_2}{2}\delta^2\right] \tag{3-11}$$

where

$$K = (1 - \alpha_1 - \alpha_2) \Big/ \left(1 + \frac{c_1 + c_2}{n_1 + n_2}\delta^2\right)$$

and

$$c_i = \frac{\alpha_i(1 - \alpha_i)}{n_i}, \qquad i = 1, 2$$

Then it can be shown that (assuming $p_1 = p_2 = \frac{1}{2}$)

$$P_{1M} = \Phi\left[\frac{-\delta(1 + \alpha_1 - \alpha_2)}{2}\right]$$

$$P_{2M} = \Phi\left[\frac{-\delta(1 - \alpha_1 + \alpha_2)}{2}\right] \tag{3-12}$$

provided that $1 - \alpha_1 - \alpha_2 > 0$, a not-unreasonable assumption (P_{1M} and P_{2M} are the error rates for the discriminant function computed from the misclassified data). In particular, if $\alpha_1 = \alpha_2$, the individual error rates are unaffected. A set of sampling experiments using $n_1 = n_2 = 11$ for $k = 2$, $n_1 = n_2 = 22$ for $k = 4$, and $n_1 = n_2 = 44$ for $k = 8$ was performed. In all experiments $\alpha_1 = 0$ and $\alpha_2 = 0$, .091, or .182. Table 3-3 gives some results from this study.

Table 3-3 Error rates for misclassified initial samples ($\alpha_1 = 0$, $\delta^2 = 1$)

α_2	$\bar{P}_1$	P_{1M}	$\bar{P}_2$	P_{2M}	k
.0	.306	.3085	.284	.3085	2
.091	.319	.3257	.276	.2935	2
.182	.343	.3409	.261	.2776	2
.0	.290	.3085	.291	.3085	4
.091	.316	.3257	.279	.2935	4
.182	.346	.3409	.255	.3776	4
.0	.311	.3085	.284	.3085	8
.091	.335	.3252	.267	.2935	8
.182	.340	.3409	.270	.2776	8

Source: Lachenbruch (*312*).

P_{1M} and P_{2M} were calculated from (3-12). The means of $\bar{P}_1$ and $\bar{P}_2$ were obtained by finding $P[D_S(\mathbf{x}) < 0 \mid \Pi_1]$ and $P[D_S(\mathbf{x}) > 0 \mid \Pi_2]$ using (2-18) and (2-19) and the known parameters for Π_1 and Π_2. Thus the behavior for moderate-sized samples was about as expected from (3-12).

Recently McLachlan (*358*) has studied the asymptotic theory for this problem. His results generally agree with those of Lachenbruch. He also notes that when $\alpha_1 = 0$, $\text{Var}(P_1)$ is an increasing function of α_2 and $\text{Var}(P_2)$ is a decreasing function of α_2.

These two papers assume that the misclassified observations are a random sample from the parent populations. This is not altogether reasonable, as the observations most likely to be misclassified initially are borderline cases. This problem has been studied by Lachenbruch (*314a*). He found that the actual error rates were relatively unaffected by nonrandom initial misclassification. However, the apparent error rates were grossly distorted and totally unreliable for any sample size.

Missing values

In practice, the robustness problems just discussed may be secondary to the problem of missing values. Many methods have been proposed, but some are too complicated for routine use. Chan and Dunn (*93*) studied the performance of various methods of handling missing values as measured by the expected actual error rate.

A. Use only complete data vectors.

B. Use all available sample values to estimate means and covariances.

BA. If method B yields a negative value of D^2, use method A; otherwise, use method B.

C. Calculate means from all available sample values, and substitute these means for missing values.

D. Obtain a regression equation for each variable, given the others, from the complete data and estimate missing values. Fill out the set of observations with one missing value by method D. Then use these for observations with two missing values, and so on.

E. Replace the complete observations by their deviations from the within-group mean divided by the standard deviation. Replace missing values by the nearest point on the first principal component and transform back to the original units.

Chan and Dunn performed a sampling experiment for a variety of correlation matrices, proportions of variables missing, distance between populations, and number of variables, k. They compared the performance of these methods to the case in which no observations were missing. For each method they calculated the probabilities of correct classification:

$$1 - P_1 = \Phi\left[\frac{[\boldsymbol{\mu}_1 - \frac{1}{2}(\hat{\boldsymbol{\mu}}_1 + \hat{\boldsymbol{\mu}}_2)]'\hat{\boldsymbol{\Sigma}}^{-1}(\hat{\boldsymbol{\mu}}_1 - \hat{\boldsymbol{\mu}}_2)}{\sqrt{(\hat{\boldsymbol{\mu}}_1 - \hat{\boldsymbol{\mu}}_2)'\hat{\boldsymbol{\Sigma}}^{-1}\boldsymbol{\Sigma}\hat{\boldsymbol{\Sigma}}^{-1}(\hat{\boldsymbol{\mu}}_1 - \hat{\boldsymbol{\mu}}_2)}}\right]$$

and

$$1 - P_2 = \Phi\left[\frac{-[\mu_2 - \frac{1}{2}(\hat{\boldsymbol{\mu}}_1 + \hat{\boldsymbol{\mu}}_2)]'\hat{\boldsymbol{\Sigma}}^{-1}(\hat{\boldsymbol{\mu}}_1 - \hat{\boldsymbol{\mu}}_2)}{(\hat{\boldsymbol{\mu}}_1 - \hat{\boldsymbol{\mu}}_2)\hat{\boldsymbol{\Sigma}}^{-1}\boldsymbol{\Sigma}\hat{\boldsymbol{\Sigma}}^{-1}(\hat{\boldsymbol{\mu}}_1 - \hat{\boldsymbol{\mu}}_2)}\right] \qquad (3\text{-}13)$$

where $\hat{\boldsymbol{\mu}}_i$ and $\hat{\boldsymbol{\Sigma}}$ are estimates of $\boldsymbol{\mu}_i$ and $\boldsymbol{\Sigma}$ by one of the methods. One hundred samples at each experimental combination were obtained to get an estimate of the expected actual correct classification rate.

Their results indicate that for $k = 2$, method D is best when $|\mathbf{R}|$ is small ($\mathbf{R}$ is the correlation matrix). Method E is best for moderate $|\mathbf{R}|$, and method C is best for large $|\mathbf{R}|$, that is, when the correlations are small. This holds true for other values of k, except that method D is never preferred for $k > 2$. As δ^2 increases or sample sizes get larger, all methods improve. If correlations are negative, the methods perform less satisfactorily.

Jackson (*265*) compared missing-values methods using real data. She points out that using only complete cases may result in considerable bias if unknowns are frequent or not randomly distributed. She considered method C, mean value replacement, and a modification of method D, an iterative procedure. She found that the iterative procedure is slightly better than the mean replacement method. This considered a single large data set and did not evaluate the performance of the methods under a variety of correlation structures as Chan and Dunn did. It is interesting to note the behavior of the two methods when real data are used.

Projects

1. Modify your program to allow you to transform variables before performing calculations. Transformations to include might be $\ln x$, $x + a$, bx, x_i/x_j, $x_i + x_j$, e^x, etc. Feel free to make up your own list. This might be done with a standard subroutine that you supply with your program, or you might require the user to write a routine that reads the data and makes the required transformations.
2. Write a routine that handles missing values by substituting the mean of the complete cases for the missing values.

4 Nonnormal and nonparametric methods

In the first three chapters we have been concerned with techniques that are optimal or asymptotically optimal for observations from normal distributions. This assumption never holds in practice, although it is often made. In some cases it may be possible to improve the performance of the linear discriminant function by the use of transformations. Other times it may be possible to use an exact distribution such as a multinomial. Nonparametric methods are a potentially valuable type of procedure, although they are not in wide use at present.

Multinomial distribution

The multinomial is an important nonnormal distribution that has been studied extensively. Suppose that there are s states or cells and that q_{ij} is the probability that an observation from Π_i falls in the jth cell, $\sum_{j=1}^{s} q_{ij} = 1, i = 1, 2.$

Then the rule is assign to Π_1 if

$$\frac{q_{1j}}{q_{2j}} > \frac{p_2}{p_1} \tag{4-1}$$

and to Π_2 otherwise. It is possible that such a rule will always assign to one population. For example, suppose that we wish to detect a diseased patient on the basis of some characteristic, such as blood group. Let the probabilities be as given in Table 4-1 (the data are artificial). If $p_1 = .2$, then the rule is: Assign to Π_1 if $q_{1j}/q_{2j} = .8/.2 = 4$, but $q_{11}/q_{21} = \frac{5}{6}$, $q_{12}/q_{22} = \frac{4}{5}$, $q_{13}/q_{23} = 2$, and $q_{14}/q_{24} = 2$, none of which is greater than 4. Thus no patient would be called diseased on the basis of this test. For this case any value of p_1 less than $\frac{1}{3}$ would produce the same result.

Two problems often appear in practice. The number of states, s, will be much larger than 4. Often the states are the set of all possible states from several unordered categorical variables. For example, the variables may be sex, race, marital status, and blood group. If these have 2, 3, 4, and 4

Table 4-1

Group	Characteristic 1	2	3	4
Π_1: diseased	.50	.20	.20	.10
Π_2: nondiseased	.60	.25	.10	.05

states, respectively, the total number of states is their product, 96. If the cell probabilities are known, a list of the states satisfying (4-1) may be made and supplied to the user. Such an assignment rule may be quite complicated; one may desire to use dummy variables and compute a linear discriminant function.

The second problem arises when the cell probabilities are unknown and must be estimated. The simplest method is to use the sample cell frequencies to estimate the parameters:

$$q_{1j} = \frac{n_{1j}}{n_{1+}} \qquad q_{2j} = \frac{n_{2j}}{n_{2+}}$$

where

$$n_{i+} = \sum_{j=1}^{s} n_{ij} \tag{4-2}$$

In general, unless the sample is large relative to s, the estimates will be unstable because a large number of parameters must be estimated. In particular, some of the n_{ij} may be zero. Some methods for overcoming this problem have been suggested by Hills (*238*) and will be discussed later.

Linhart (*332*) studied the problem of classifying multinomial data and suggested use of the likelihood ratio rule, (4-1), but did not comment on the problems associated with estimating cell probabilities for large numbers of cells.

Cochran and Hopkins (*115*) discussed three basic problems in the use of the multinomial distribution with estimated cell probabilities:

1. The effect of the initial sample sizes on the performance of the proposed rule.
2. The relative discriminating power of qualitative and continuous variates.
3. Use of classification experience for improvement of the rule.

The optimum rule is the one based on the likelihood ratio, and it is estimated by the ratio of observed cell frequencies. That is, assign to Π_1 if $n_{1j}/n_{2j} > p_2/p_1$ when the unknown observation falls in state j. They noted that (1) the estimated probability of misclassification has an approximately binomial sampling error, (2) the actual error rate is greater than the optimum, but (3) the apparent error rate is less than the optimum.

Sometimes it may be useful to replace continuous variables by several categories because the assignment rule becomes too complicated to be practical. Cochran and Hopkins studied the effect of partitioning a continuous variable into k' categories. If there are k continuous variables, there will be $(k')^k$ multinomial states. For example, if we have two variables (k) with four states each (k'), an observation can fall into one of $(4)^2 = 16$ possible states. For various values of k and k', under the assumptions that each of the k continuous variables was normal and independent of the others, and $p_1 = p_2$, they were able to calculate the relative discriminating power of the categorized variables to the continuous ones. Table 4-2, which gives their results, indicates that for two categories with an odd value of k, the relative power is greater than the asymptotic power. This peculiarity arises because when $k = 1$, the assignment rule is the same as the optimum one, so the relative power is 1. This will also occur when $k' = 4$, $k = 1$.

In general, the greater the number, k', of qualitative states, the closer the multinomial rule comes to the optimum likelihood ratio for the normal variables. However, in practical situations improvement in the assignment rule must be weighed against the disadvantage of having many probabilities to estimate.

Hills (*238*) was concerned with the problem of estimating likelihood ratios for Bernoulli variables. As there are 2^k possible outcomes, the distribution may be considered multinomial, although in some cases it is possible to represent the distribution more economically (see Chapter 3). He proposed the use of nearest-neighbor procedures to help overcome the

Table 4-2 *Relative discriminating power of qualitative versus continuous normal variates*

k':	2		3		4		5	
k:	2	∞	2	∞	2	∞	2	∞
	.50	.64	.74	.81	.81	.88	.89	.92

k':	2							3				
k:	2	3	4	5	6	7	∞	2	3	4	5	∞
	.50	.74	.56	.70	.58	.68	.64	.74	.76	.77	.78	.8

Source: Cochran and Hopkins (*115*).

problem of small (or zero) cell frequencies. For the Bernoulli case, a cell may be represented by the corresponding pattern of zeros and ones. For example, with $k = 3$ the possible patterns are

$$\begin{array}{cccc} 000 & 001 & 010 & 011 \\ 100 & 101 & 110 & 111 \end{array} \tag{4-3}$$

A near neighbor of order 1 is one that differs from the pattern in only one variable. The near neighbors of 010 are

$$011 \quad 000 \quad 110 \tag{4-4}$$

If the cell count for the jth cell is n_{ij}, then the nearest-neighbor procedure assigns the observation to Π_1 if

$$\frac{(n_{1j} + \sum_A n_{1j})/n_1}{(n_{2j} + \sum_A n_{2j})/n_2} > \frac{p_2}{p_1} \tag{4-5}$$

where A is the set of near neighbors of state j. Hills comments that the estimate of the likelihood ratio has less sampling variability than the simple method using cell frequencies. Various adjustments to the rule are discussed also. When defining near neighbors one may use only those variables that are unimportant (on a priori grounds) for discriminating, or one may use near neighbors of order 2 or more.

Hills also notes some problems with the use of the error rates as the criterion for evaluating an assignment rule and proposes

$$\sum_{j=1}^{s} (q_{1j} - q_{2j}) \ln \frac{q_{1j}}{q_{2j}}$$

and (4-6)

$$\sum_{j=1}^{s} \frac{(q_{1j} - q_{2j})^2}{q_{1j} + q_{2j}}$$

as alternative measures.

In general, work with discrete data requires care on the part of the statistician and his client. Programs to do the computations are not generally available. Fortunately, there is evidence that the linear discriminant function classifies observations satisfactorily in many situations. However, an unordered categorical variable (for example, marital status) must never be used in its original form in a program that computes a linear discriminant function. It may be converted to Bernoulli form by means of dummy variables. For example,

$$X_1 = 1 \text{ if single, 0 otherwise}$$

$$X_2 = 1 \text{ if married, 0 otherwise}$$

$$X_3 = 1 \text{ if divorced, 0 otherwise}$$

may be used in place of the variable Y = marital status (0 = single, 1 = married, 2 = divorced, 3 = widowed). In practice there may be "unknown" and "other" codes that must be considered.

Other nonnormal distributions

Very little has been done in discrimination between populations with distributions other than the normal or the multinomial. In one sense the multinomial is completely adequate for discrete distributions if the parameters are known. However, if they must be estimated, it is better to use the appropriate distribution, as the number of parameters to be estimated cannot be greater than the number of cells. For continuous distributions, essentially no multivariate distributions other than the normal are widely known. No examples of discrimination using a continuous nonnormal multivariate distribution have appeared in the literature. For univariate nonnormals, the theory is straightforward: Use $f_1(x)/f_2(x)$ or its logarithm and assign to Π_1 or Π_2 on that basis. For example, the exponential family of distributions has the form

$$f(x \mid \boldsymbol{\theta}_i) = H(x)G(\boldsymbol{\theta}_i) \exp\left[\sum_{j=1}^{r} \mu_j(x)\phi_j(\boldsymbol{\theta}_i)\right] \tag{4-7}$$

This family includes a wide variety of common distributions such as the normal, exponential, Poisson, and binomial. For this family,

$$\frac{f_1(x)}{f_2(x)} = \frac{f(x \mid \boldsymbol{\theta}_1)}{f(x \mid \boldsymbol{\theta}_2)} = \frac{G(\boldsymbol{\theta}_1)}{G(\boldsymbol{\theta}_2)} \exp\left\{\sum_{j=1}^{r} \mu_j(x)[\phi_j(\boldsymbol{\theta}_1) - \phi_j(\boldsymbol{\theta}_2)]\right\} \tag{4-8a}$$

and

$$\ln \frac{f_1(x)}{f_2(x)} = \sum_{j=1}^{r} \mu_j(x)[\phi_j(\boldsymbol{\theta}_1) - \phi_j(\boldsymbol{\theta}_2)] + C(\boldsymbol{\theta}_1, \boldsymbol{\theta}_2) \tag{4-8b}$$

In many cases this leads to very simple classification regions.

Day and Kerridge (*146*) suggested an approach that allows considerable latitude in choice of distributions, yet remains quite simple to apply. Suppose that

$$f_i(\mathbf{x}) = c_i \exp[-\tfrac{1}{2}(\mathbf{x} - \boldsymbol{\mu}_i)'\boldsymbol{\Sigma}^{-1}(\mathbf{x} - \boldsymbol{\mu}_i)]h(\mathbf{x}) \tag{4-9}$$

where c_i is a constant chosen to make the total probability mass equal to 1, and $h(\mathbf{x})$ is an arbitrary function of $\mathbf{x}$ which is integrable and nonnegative. A wide variety of special cases may be found under this model. If $h(\mathbf{x})$ is identically equal to 1, the multivariate normal case holds. If $h(\mathbf{x}) = 1$ when each component of $\mathbf{x}$ is 1 or 0 and $\mathbf{\Sigma} = \mathbf{I}$, the independent Bernoulli variate case is obtained. If $\mathbf{\Sigma} \neq \mathbf{I}$, this yields correlated Bernoulli variates. To obtain a mixed discrete-continuous model, $h(\mathbf{x}) = 1$ if a subset of the components of $\mathbf{x}$ has specified integral values. Finally, it was noted that in many biomedical problems one group is fairly symmetrical (usually the normals) whereas the other is skewed. This can be handled by having $h(\mathbf{x})$ constant near the mean of the first group, and increasing in the region beyond the mean of the second. It was pointed out that the covariance matrix of $\mathbf{x}$ is equal to $\mathbf{\Sigma}$ if $h(\mathbf{x}) \equiv 1$, but not if $h(\mathbf{x}) \not\equiv 1$.

When we use Bayes theorem we obtain

$$P(\Pi_1 \mid \mathbf{x}) = \frac{p_1c_1 \exp[-\frac{1}{2}(\mathbf{x} - \boldsymbol{\mu}_1)'\mathbf{\Sigma}^{-1}(\mathbf{x} - \boldsymbol{\mu}_1)]}{p_1c_1 \exp[-\frac{1}{2}(\mathbf{x} - \boldsymbol{\mu}_1)'\mathbf{\Sigma}^{-1}(\mathbf{x} - \boldsymbol{\mu}_1)] + p_2c_2 \exp[-\frac{1}{2}(\mathbf{x} - \boldsymbol{\mu}_2)'\mathbf{\Sigma}^{-1}(\mathbf{x} - \boldsymbol{\mu}_2)]}$$

$$= \frac{\exp(\mathbf{x}'\mathbf{b} + A)}{1 + \exp(\mathbf{x}'\mathbf{b} + A)} \qquad (4\text{-}10)$$

where $\mathbf{b} = \mathbf{\Sigma}^{-1}(\boldsymbol{\mu}_1 - \boldsymbol{\mu}_2)$ and

$$A = -\tfrac{1}{2}(\boldsymbol{\mu}_1 + \boldsymbol{\mu}_2)'\mathbf{\Sigma}^{-1}(\boldsymbol{\mu}_1 - \boldsymbol{\mu}_2) + \ln \frac{p_1c_1}{p_2c_2}$$

In general, one must use an iterative procedure to find $\mathbf{b}$ and A since $\bar{\mathbf{x}}_1$, $\bar{\mathbf{x}}_2$, and $\mathbf{S}$ are not unbiased estimates of $\boldsymbol{\mu}_1$, $\boldsymbol{\mu}_2$, and $\mathbf{\Sigma}$ and because the c_i terms depend on the $h(\mathbf{x})$ function. Individuals are assigned to Π_1 if $P(\Pi_1 \mid \mathbf{x}) > .5$ and to Π_2 otherwise. Day and Kerridge indicate that they were able to classify somewhat better using an iterative procedure than with the discriminant coefficients. In any case, $\mathbf{S}^{-1}(\bar{\mathbf{x}}_1 - \bar{\mathbf{x}}_2)$ should be a reasonable starting value for the iterative procedure. Further discussion of this model appears in Chapter 6.

Nonparametric rules

Thus far we have considered models in which the distributions were specified either completely or to the point at which parameters could be estimated and assignment rules developed. Another important situation is the one in which the form of the distribution is not specified. In this case one is led to nonparametric or distribution-free rules. To date, no methods are widely available, and the user of discriminant analysis must develop the rule himself. It is thus a strong temptation to simply use a packaged program such as BMD 07M.

In general, the rules obtain estimates of the density function for Π_1 and Π_2 at the point to be classified and form the ratio of the density estimates. The first method proposed was the nearest-neighbor rule of Fix and Hodges (*175*). The nonparametric estimate of $f_1(\mathbf{x})/f_2(\mathbf{x})$ is found as follows. Let $\mathbf{x}_{11}\cdots\mathbf{x}_{1n_1}$ be a sample from Π_1 and $\mathbf{x}_{21}\cdots\mathbf{x}_{2n_2}$ be a sample from Π_2 and let $\mathbf{x}$ be the observation to be assigned to Π_1 or Π_2. Use some distance, function $d(\mathbf{x}_{ij}, \mathbf{x})$ and order the values $d(\mathbf{x}_{ij}, \mathbf{x})$. Choose some integer, K, and let K_i be the number of observations from Π_i in the K closest observations to $\mathbf{x}$. Then assign $\mathbf{x}$ to Π_1 if

$$\frac{K_1}{n_1} > \frac{K_2}{n_2} \tag{4-11}$$

A simple generalization of the rule to account for unequal a priori probabilities is to assign to Π_1 if

$$\frac{K_1}{n_1} \bigg/ \frac{K_2}{n_2} > \frac{p_2}{p_1} \tag{4-12}$$

These estimates are consistent and the error rate tends to the error rate of the maximum likelihood rule when $n_i \to \infty$. This method has been studied by Cover and Hart (*129*), Peterson (*413*), Patrick (*407*), Patrick and Fisher (*409*), Pelto (*410*), and Palmershein (*405*), and various modifications have been suggested.

More recently, various methods of estimating cumulatives and densities have been proposed. Most of these have been concerned with the univariate problem, but some do consider multivariate estimation. As might be expected, they require larger samples than is needed for parameter estimation in known distributions. A few of these methods will be mentioned here. The interested reader may refer to Das Gupta (*142a*) or Wegman (*563*) for further information. Relatively little has been done in applying these estimates to discriminant problems.

Parzen (*406*) proposed a class of estimators of the form

$$\widehat{f(x)} = \frac{1}{nh} \Sigma K\left(\frac{x - x_i}{h}\right) \tag{4-13}$$

where x_i are observed data points and $K(x)$ and $h(n)$ are functions satisfying:

1. $\int_{-\infty}^{\infty} K(z)\, dz = 1$

2. $\int_{-\infty}^{\infty} |K(z)|\, dz < \infty$

3. $\sup |K(z)| < \infty$

4. $\lim_{z\to\infty} |zK(z)| = 0$

5. $\lim_{n\to\infty} h(n) = 0$

6. $\lim_{n\to\infty} nh(n) = \infty$

The estimates of $f(x)$ are consistent and asymptotically normal. A wide variety of functions $K(z)$ will satisfy conditions 1–4. The approach was extended to multivariate distributions by Cacoullos (*85a*).

Specht (*506*) considered estimates of densities using

$$K\left(\frac{x}{\sigma}\right) = \frac{1}{\sqrt{2\pi}} e^{-1/2(x/\sigma)^2} \tag{4-14}$$

He suggested using truncated series expansions of the exponential functions involved so that the density estimator could be evaluated for any point, x, without having to retain all the sample observations. He discussed the order of truncation and the selection of the smoothing parameter σ, and he showed that the optimum σ decreases with sample size and that a moderate number of terms will be sufficient for fairly accurate estimation near the center of the distribution. In an earlier paper (*504*) he gave a technique for generating polynomial discriminant functions based on multivariate expansions of this type. He also applied it to the problem of diagnosis of vectorcardiograms (*503*). This method involves computing a set of rather complicated coefficients and assigning to Π_i on the basis of the ratio of two messy polynomials. Because of this, a computer is required for this technique.

Another method of estimating densities is to use Fourier series expansions [Kronmal and Tarter (*303*), Tarter and Kronmal (*522*), Tarter and Raman (*523*)]. These estimates have the form

$$\hat{f}_M(\mathbf{x}) = H(\mathbf{x}) + \sum_{\mathbf{r}\in M} \hat{B}_{\mathbf{r}} e^{2\pi i \mathbf{r}'\mathbf{x}} \tag{4-15}$$

where M is a set of k-tuples of integers and $\hat{B}_{\mathbf{r}} = 1/n \sum_{j=1}^{n} e^{-2\pi i \mathbf{r}'\mathbf{x}_j}$. It was shown that the number of terms needed in the estimate is less than 15 for a variety of univariate distributions and sample sizes (*523*). Tarter and Raman (*523*) indicate that satisfactory estimates of $f(\mathbf{x})$ may be had with as few as 45 observations in the bivariate situation.

Procedures based on statistically equivalent blocks have been proposed by Gessaman (*201*) and by Anderson (*22*). A simple procedure of this type is as follows. Order the sample from Π_1 on the basis of, say, the first variable. Partition the sample into l equal groups. Now order each of these groups on the basis of the second variable and partition them into l groups. Do this for all k variables. Then there will be an approximately equal number in each block. Determine the number of the sample of Π_2 falling in each block. Suppose that n_{1i} is the number in the ith cell from Π_1 and n_{2i} the number in the ith cell from Π_2. Then assign to Π_1 if

$$\frac{n_{1i}/n_1}{n_{2i}/n_2} > \frac{p_2}{p_1} \tag{4-16}$$

This procedure can also be carried out using the sample from Π_2 to construct the blocks. In general, fairly large samples are needed for this procedure unless the number of variables is restricted. It is possible to use a subset of the k variables for constructing the blocks or to use a linear combination of variables such as principal components. This is a nearest-neighbor type of procedure.

Gessaman and Gessaman (*202*) compared the performance of this procedure with the linear discriminant function and three nonparametric density estimator techniques. They used $k = 2$ and three pairs of populations. Each of the first pair of populations was normal with different means and different covariances. Each of the second pair had the same means but different covariances, and the third pair had different means but the same covariances. Samples of size $n_1 = n_2 = 729$, 200, or 64 were generated, and the procedures were evaluated by observing the error rates on 500 additional observations, 250 from Π_1, and 250 from Π_2. The nonparametric density estimates were (1) a Parzen estimator (*406*), (2) the Quesenberry–Loftsgaarden estimator (*424*), and (3) the Gessaman estimator based on statistically equivalent blocks (*201*). The results of this experiment are given in Table 4-3. The linear discriminant function is poor for population pairs I and II because of the unequal covariance structure. However, its performance is quite good on the third pair. The Parzen estimator is very slightly better than the Quesenberry–Loftsgaarden estimator. The nearest-neighbor procedure does quite well in all situations.

One problem with some nonparametric estimators is that they require that the entire initial sample be retained. This is a considerable hindrance if one is attempting to develop a practicable assignment procedure. Some of the methods proposed have overcome this problem by defining the assign-

Table 4-3 Error rates of five procedures

Procedure	Population I			Population II			Population III		
$n_1 = n_2$	729	200	64	729	200	64	729	200	64
Discriminant function	.116	.118	.112	.502	.496	.498	.150	.158	.166
Parzen	.090	.100	.100	.328	.332	.354	.172	.168	.176
Loftsgaarden–Quesenberry	.096	.108	.102	.320	.336	.370	.160	.162	.178
Gessaman blocks	.110	.108	.140	.360	.356	.436	.160	.160	.200
Gessaman nearest-neighbor	.090	.096	.098	.246	.292	.304	.142	.158	.178

Source: Gessaman and Gessaman (*202*).

ment regions nonparametrically or by developing the rules in polynomial or series form. An additional problem is the fairly large number of parameters to be estimated. However, the methods seem to work satisfactorily with moderate sample sizes.

The development of these methods for discrimination is still in its early stages. Problems involving selection of variables have not been studied in this context, and it is still hard to approximate multivariate distributions of high dimensionality.

Example

An example in which these methods have been used involves the vectorcardiogram, which has been used successfully in the diagnosis of heart disease. This device is similar to the electrocardiogram except that three leads are used instead of twelve. The leads are placed so that they form approximately an orthogonal set, and the outputs are recorded simultaneously rather than sequentially; thus phase information can be obtained. By converting the analog signals to digital form, discriminant analysis may be performed on the data. Specht (*503*) used the QRS complex to classify patients into normal or abnormal groups by sampling every 5 milliseconds up to 75 ms after the onset of the QRS phase. In addition, the duration of the QRS was recorded giving a total of 46 measurements from the three leads.

The distribution of these measures cannot be considered multivariate normal, so he used a density-approximation technique. In Π_i, with n_i data points, it is possible to approximate $f_i(\mathbf{x})$ at the unknown point $\mathbf{x}$ by

$$\widehat{f_i(\mathbf{x})} = \frac{1}{n_i} \sum_{j=1}^{n_i} g(\mathbf{x} - \mathbf{x}_j) \tag{4-17}$$

where

$$g(\mathbf{x} - \mathbf{x}_i) = \frac{1}{(2\pi)^{k/2}\sigma^k} \exp\left[\frac{(\mathbf{x} - \mathbf{x}_j)'(\mathbf{x} - \mathbf{x}_j)}{2\sigma^2}\right] \tag{4-18}$$

and σ^2 is a smoothing parameter chosen by the investigators. It is easy to see that (4-18) can be written as

$$\widehat{f_i(\mathbf{x})} = \frac{1}{n_i(2\pi)^{k/2}\sigma^k} \exp\left(-\frac{\mathbf{x}'\mathbf{x}}{2\sigma^2}\right) \sum_{j=1}^{n_i} \exp\left(\frac{\mathbf{x}_j'\mathbf{x} + B_j}{\sigma^2}\right) \tag{4-19}$$

where $B_j = -\mathbf{x}_j'\mathbf{x}_j/2$. By expanding $\exp(\mathbf{x}_j'\mathbf{x}/\sigma^2)$ in a Taylor series, the density (4-19) can be approximated by a function of the form

$$\widehat{f_i(\mathbf{x})} = \frac{1}{\sigma^k(2\pi)^{k/2}} \exp\left(-\frac{\mathbf{x}'\mathbf{x}}{2\sigma^2}\right) [\sum D^{(i)}_{j1 \cdots jk},\, x_1{}^{j1} \cdots x_k{}^{jk}] \tag{4-20}$$

where

$$D^i_{j1 \cdots jk} = \frac{1}{\sigma^{2h} j_1! \cdots j_k!} \frac{1}{n_i} \sum_{j=1}^{n_i} X^{j_i}_{j1} \cdots X^{jk}_{jk} \exp\left(\frac{B_j}{\sigma^2}\right)$$

and $h = j_1 + \cdots + j_k$. Often, relatively few of these terms are sufficient to approximate the distribution.

The data consisted of 224 normals and 88 abnormals. Because the ECG waveform has different properties for males and females, and because it varies with age, all subjects were female between 21 and 50. The data were partitioned into an initial sample consisting of 192 normals and 57 abnormals and a testing set of 32 normals and 31 abnormals. Specht, instead of minimizing total error rate by choosing $(1 - p)/p$ as the cutoff, chose to minimize the false negative rate subject to the condition that the false positive rate was .05. For $\sigma = 4$, he found (1) using the leaving-one-out method there were 95 percent correct on the normal patients and 86 percent correct on the abnormals, and (2) in the testing sample 97 percent correct on the normals and 90 percent on abnormals. This was obtained with 30 coefficients. Compared with clinical diagnosis based on ECG alone, he found that clinicians detected 53 percent of abnormal patients compared with a 90 percent performance for the polynomial rule.

This method has certain limitations. First, fairly large samples are needed, as with any nonparametric method. The 30 coefficients derived were chosen from 369 coefficients. Second, the coefficients estimated from high powers of the variables are unstable. It would have been interesting if the investigator had compared this method to other nonparametric rules and the linear discriminant function.

Projects

1. Write a program that will do multinomial discrimination using the maximum likelihood procedure and the nearest-neighbor procedure.
2. Write a program that will handle multinomial discrimination when it is known that certain subsets of variables are known to be independent, that is, $f_i(\mathbf{x}) = f_{1i}(\mathbf{x}_1)f_{2i}(\mathbf{x}_2)\cdots f_{si}(\mathbf{x}_s)$.

5 *Multiple-group problems*

This far we have been concerned primarily with the problem of assigning an individual to one of two groups. The multiple-group problem has not been as extensively studied as the two-group problem, for two main reasons. First, the essence of the problem is often contained in the two-group case. For example, estimation of error rates is essentially similar for two groups or for g groups, though a complicating feature in the second case is the possibility of $g - 1$ types of error. Second is that the multiple-group case involves more complex sampling situations. In discriminant analysis, much of the research has been done by use of sampling experiments. For two groups it is possible to arrange the group means on a straight line, with one mean at the origin. For multiple groups the many possible configurations of means lead to problems in the design of sampling experiments. The theoretical solution to the optimal decision rule which minimizes the total error is not difficult to obtain. However, in practice, questions about the robustness of this solution to insults such as nonnormality and unequal covariances remain unanswered. A second solution using canonical variables is a direct generalization of Fisher's approach. The same robustness questions remain open for any method that employs canonical variables.

Optimal classification rule

Suppose that there are g groups with probability density functions $f_i(\mathbf{x})$, $i = 1, \ldots, g$, and that we wish to assign an unknown observation, $\mathbf{x}$, to one of them. If we assign $\mathbf{x}$ to Π_i if $\mathbf{x}$ falls in some region R_i, we can define the probabilities of misclassification of an observations from Π_i into Π_j as

$$P(j \mid i) = \int_{R_j} f_i(\mathbf{x})\, d\mathbf{x} \tag{5-1}$$

For a particular $\mathbf{x}$ we have the conditional probability of Π_i given $\mathbf{x}$ as

$$P(\Pi_i \mid \mathbf{x}) = \frac{p_i f_i(\mathbf{x})}{\sum_{l=1}^{g} p_l f_l(\mathbf{x})} \tag{5-2}$$

If we assign $\mathbf{x}$ to Π_j the expected loss is

$$\frac{\sum_{i \neq j} p_i f_i(\mathbf{x}) c_{ji}}{\sum_{l=1}^{g} p_l f_l(\mathbf{x})} \tag{5-3}$$

where c_{ji} is the cost of misclassification of an observation from Π_i into Π_j and $c_{ij} = 0$. To minimize the expected loss we must minimize the numerator of (5-3). Now the loss due to assigning Π_j is less than that due to assigning to Π_m if

$$\sum_{i \neq j} p_i f_i(\mathbf{x}) c_{ji} < \sum_{i \neq m} p_i f_i(\mathbf{x}) c_{mi} \tag{5-4}$$

If $c_{ij} = c$, for all i, j, (5-4) is equivalent to

$$p_m f_m(\mathbf{x}) < p_j f_j(\mathbf{x}) \tag{5-5}$$

Suppose that we wish to discriminate among dysplasia (Π_1), in situ cancer (Π_2), and invasive cancer (Π_3) and somehow have been able to postulate costs of making errors as shown in Table 5-1. Suppose that $p_1 = .8$, $p_2 = .19$, and $p_3 = .01$. Consider three patients whose $\mathbf{x}$ measurements have corresponding f_i values, as shown in Table 5-2. The values of $\sum_{i \neq j} p_i f_i c_{ji}$ for patient 1 are

$$p_2 f_2 c_{21} + p_3 f_3 c_{31} = (.19)(.2)50 + (.01)(.02)(1000) = 2.10$$

$$p_1 f_1 c_{12} + p_3 f_3 c_{32} = (.8)(2)(1) + (.01)(.02)(50) = 1.61$$

$$p_1 f_1 c_{13} + p_2 f_2 c_{23} = (.8)(2)(100) + (.19)(.2)(100) = 163.80$$

So patient 1 would be assigned to Π_2, the in situ group. The corresponding values for patient 2 are .06, 4.005, and 401.9, so she would be assigned to Π_1, the dysplasia group. The values for patient 3 are 100.95, 5.08, and 2.70, and she would be assigned to Π_3, the invasive group.

Table 5-1

	Assign to		
True group	*Dysplasia*	*In situ*	*Invasive*
Dysplasia	0	1	100
In situ	50	0	100
Invasive	1000	50	0

Table 5-2

Patient	f_1	f_2	f_3
1	2	.2	.02
2	5	.1	.01
3	.01	.1	10

If (5-4) holds for all $m \neq j$, then we assign $\mathbf{x}$ to Π_j. For the particular case $c_{ji} = 1$ if $i \neq j$ and 0 if $i = j$, the rule is assign to Π_j if $p_j f_j(\mathbf{x}) > p_m f_m(\mathbf{x})$ for all $m \neq j$. Anderson (*20*) shows that this rule is admissible and that it is Bayes. For the special case when the distributions are multivariate normal and $c_{ji} = 1$ if $i \neq j$, 0 if $i = j$, mean $\boldsymbol{\mu}_i$, and covariance $\boldsymbol{\Sigma}_i$, the rule is assign to Π_i if

$$p_i \frac{1}{(2\pi)^{k/2} \mid \boldsymbol{\Sigma}_i \mid^{1/2}} \exp[-\tfrac{1}{2}(\mathbf{x} - \boldsymbol{\mu}_i)'\boldsymbol{\Sigma}_i^{-1}(\mathbf{x} - \boldsymbol{\mu}_i)]$$

$$= \max_j p_j \frac{1}{(2\pi)^{k/2} \mid \boldsymbol{\Sigma}_j \mid^{1/2}} \exp[-\tfrac{1}{2}(\mathbf{x} - \boldsymbol{\mu}_j)'\boldsymbol{\Sigma}_j^{-1}(\mathbf{x} - \boldsymbol{\mu}_j)] \quad (5\text{-}6)$$

Taking logarithms the rule is assign to Π_i if

$$\ln p_i - \tfrac{1}{2} \ln \mid \boldsymbol{\Sigma}_i \mid - \tfrac{1}{2}(\mathbf{x} - \boldsymbol{\mu}_i)'\boldsymbol{\Sigma}_i^{-1}(\mathbf{x} - \boldsymbol{\mu}_i)$$

$$= \max_j \{\ln p_j - \tfrac{1}{2} \ln \mid \boldsymbol{\Sigma}_j \mid - \tfrac{1}{2}(\mathbf{x} - \boldsymbol{\mu}_j)'\boldsymbol{\Sigma}_j^{-1}(\mathbf{x} - \boldsymbol{\mu}_j)\} \quad (5\text{-}7)$$

If $\boldsymbol{\Sigma}_i = \boldsymbol{\Sigma}$ for $i = 1, \ldots, g$, the rule becomes assign to Π_i if

$$\ln p_i + \left(\mathbf{x} - \frac{\boldsymbol{\mu}_i}{2}\right)' \boldsymbol{\Sigma}^{-1}\boldsymbol{\mu}_i = \max_j \left[\ln p_j + \left(\mathbf{x} - \frac{\boldsymbol{\mu}_j}{2}\right)' \boldsymbol{\Sigma}^{-1}\boldsymbol{\mu}_j\right] \quad (5\text{-}8)$$

Most widely available computer programs are based on (5-8). They are thus subject to possible nonrobustness because of unequal covariance matrices or nonnormality.

If the parameters are unknown, plug-in estimates may be used. For estimating $\boldsymbol{\Sigma}$, the pooled sample covariance may be used:

$$\hat{\boldsymbol{\Sigma}} = \mathbf{S} = \frac{1}{\Sigma(n_{\cdot} - 1)} \sum_{i=1}^{g} \sum_{j=1}^{n_i} (\bar{\mathbf{x}}_{ij} - \bar{\mathbf{x}}_{i\cdot})(\bar{\mathbf{x}}_{ij} - \bar{\mathbf{x}}_{i\cdot})' \quad (5\text{-}9)$$

For estimating $\boldsymbol{\mu}_i$, the sample means are used:

$$\hat{\boldsymbol{\mu}}_i = \bar{\mathbf{x}}_{i\cdot} = \frac{1}{n_i} \sum_{i=1}^{n_i} \mathbf{x}_{ij} \quad (5\text{-}10)$$

These have been shown to be asymptotically optimal.

Canonical vectors

The second approach to the multiple-group problem is based on Fisher's original method. It develops canonical variates based on the between-group and within-group covariance matrices.

Let B be the between-groups covariance matrix and W the within-groups covariance matrix. Then, if the parameters are known, we have

$$\mathbf{B} = \frac{1}{g} \sum_{i=1}^{g} (\boldsymbol{\mu}_i - \bar{\boldsymbol{\mu}})(\boldsymbol{\mu}_i - \bar{\boldsymbol{\mu}})' \tag{5-11}$$

and

$$\mathbf{W} = \boldsymbol{\Sigma}$$

where $\bar{\boldsymbol{\mu}} = 1/g \sum_{i=1}^{g} \boldsymbol{\mu}_i$ and $\boldsymbol{\Sigma}$ is the population covariance matrix. If the parameters are unknown,

$$\hat{\mathbf{B}} = \frac{1}{g} \sum_{i=1}^{g} (\bar{\mathbf{x}}_{i.} - \bar{\mathbf{x}}_{..})(\bar{\mathbf{x}}_{i.} - \bar{\mathbf{x}}_{..})'$$

$$\hat{\mathbf{W}} = \frac{1}{\Sigma n_i - g} \sum_{i=1}^{g} \sum_{j=1}^{n_i} (\mathbf{x}_{ij} - \bar{\mathbf{x}}_{i.})(\mathbf{x}_{ij} - \bar{\mathbf{x}}_{i.})' \tag{5-12}$$

and

$$\bar{\mathbf{x}}_{..} = \frac{1}{\Sigma n_i} \sum_{i=1}^{g} \sum_{j=1}^{n_i} \mathbf{x}_{ij}.$$

In the original article (*172*), Fisher suggested finding the linear compound, $\boldsymbol{\lambda}$, which maximized

$$\gamma = \frac{\boldsymbol{\lambda}'\mathbf{B}\boldsymbol{\lambda}}{\boldsymbol{\lambda}'\mathbf{W}\boldsymbol{\lambda}} \tag{5-13}$$

For two groups ($g = 2$) this reduces to the usual linear discriminant function. Now

$$\frac{\partial \gamma}{\partial \boldsymbol{\lambda}} = \frac{2\mathbf{B}\boldsymbol{\lambda}(\boldsymbol{\lambda}\mathbf{W}\boldsymbol{\lambda}) - 2\mathbf{W}\boldsymbol{\lambda}(\boldsymbol{\lambda}'\mathbf{B}\boldsymbol{\lambda})}{(\boldsymbol{\lambda}'\mathbf{W}\boldsymbol{\lambda})^2} = 0 \tag{5-14}$$

This yields

$$\mathbf{B}\boldsymbol{\lambda} - \gamma\mathbf{W}\boldsymbol{\lambda} = 0 \tag{5-15}$$

or

$$(\mathbf{B} - \gamma\mathbf{W})\boldsymbol{\lambda} = 0$$

This equation has a nontrivial solution only if

$$|\mathbf{B} - \gamma \mathbf{W}| = 0 \tag{5-16}$$

The solutions to this equation are the eigenvalues of $\mathbf{W}^{-1}\mathbf{B}$. There are no more than $\min(g-1, k)$ nonzero solutions. The corresponding eigenvectors are the linear compounds $\boldsymbol{\lambda}$ that will be used for discriminating.

If the observations are normal, then $\mathbf{y} = \boldsymbol{\lambda}'\mathbf{x}$ is normal also. If several eigenvectors are used, they are independent. By suitable rescaling it can be shown that if r vectors are used, the rule becomes: Assign to Π_i if

$$\sum_{l=1}^{r} [\boldsymbol{\lambda}_l'(\mathbf{x} - \boldsymbol{\mu}_i)]^2 = \min_j \sum_{l=1}^{r} [\boldsymbol{\lambda}_l'(\mathbf{x} - \boldsymbol{\mu}_j)]^2 \tag{5-17}$$

or

$$(\mathbf{y} - \tfrac{1}{2}\mathbf{v}_i)'\mathbf{v}_i = \max_j (\mathbf{y} - \tfrac{1}{2}\mathbf{v}_j)'\mathbf{v}_j$$

where $\mathbf{y}' = (\boldsymbol{\lambda}_1'\mathbf{x}, \boldsymbol{\lambda}_2'\mathbf{x}, \ldots, \boldsymbol{\lambda}_r'\mathbf{x})$ and $\mathbf{v}_i' = (\boldsymbol{\lambda}_1'\boldsymbol{\mu}_i, \ldots, \boldsymbol{\lambda}_r'\boldsymbol{\mu}_i)$. One would be interested in using this canonical-vector approach for several reasons. First, if k is quite large compared to g, a convenient representation of the information may be made. Second, if the means lie in a fairly low-dimensional subspace of the sample space, the canonical vectors will be useful. In particular, if the means are collinear, a single vector will be sufficient for discrimination. Third, plotting the first few canonical vectors can be useful in exploring the data. The BMD 07M program has an option for plotting the canonical variables. Fourth, the eigenvalues can be used to test the hypothesis that the means are equal. Possible disadvantages of this method lie in the known suboptimality of this solution for normal variables if all vectors are not used.

When the assumptions of common covariances and normality are violated, problems arise in the use of either method.

Comparison of methods

Lachenbruch (*317*) compared the performance of the two methods for two extreme cases when the assumptions of normality and common covariances held. It was assumed that $p_i = 1/g$, $i = 1, \ldots, g$. The cases concerned the configuration of population means. The most favorable case for the canonical vector method is when the means are collinear. The least favorable case arises when they are arranged in a regular simplex. For $g = 3$ this is an equilateral triangle, and for $g = 4$, it is a tetrahedron.

If the means are collinear and equally spaced, we may assume that $\boldsymbol{\mu}_i' = (i-1)\delta, 0, \ldots, 0$. Then it can be shown that the optimal rule is assign to

$$\Pi_1 \text{ if } -\infty < x_1 < \frac{\delta}{2}$$

$$\Pi_i \text{ if } \frac{\delta}{2} + \delta(i-2) < x_1 < \frac{\delta}{2} + \delta(i-1) \qquad i = 2, \ldots, g-1$$

$$\Pi_g \text{ if } \frac{\delta}{2} + \delta(g-2) < x_1 < \infty$$

In this situation the probabilities of correct classification for Π_1 and Π_g are equal to $\Phi(\delta/2)$ and the probabilities of correct classification for $\Pi_2, \ldots, \Pi_{g-1}$ are equal to $\Phi(\delta/2) - \Phi(-\delta/2)$. For the simplex case, the probability of correct classification can be shown to be

$$\int_{-\infty}^{\infty} \left[\Phi\left(\frac{\delta^2/2 - y}{\sqrt{\delta^2/2}}\right)\right]^{g-1} \phi(y)\, dy \tag{5-18}$$

where δ^2 is the pairwise distance between vertices of the simplex.

Sampling experiments were performed to compare the behavior of the two methods under the two extreme conditions. It was found that:

1. Eigenvectors performed much better on collinear configurations than on simplexes. A single canonical vector did almost as well as the optimal rule for collinear data.

2. Increasing sample size decreases the apparent probability of correct classification. This is as expected, since the known bias is being reduced.

3. Increased number of groups reduced the probability of correct classification because there are more chances for erroneous assignment.

4. Increased number of variables increased the apparent correct classification rate. This is more pronounced for the plug-in optimal rule than for the canonical-vector method.

Michaelis (*371*) studied multiple-group discrimination using populations with unequal covariance matrices. He compared the performance of the functions based on the equal covariance assumption with the performance of the function when the covariance matrices were unequal. The means and covariance matrices used were obtained from ECG data obtained on four groups of patients and a "normal" group (that is, $g = 5$). The simulated data used in the sampling experiment were generated as 8-variate multivariate normal with the appropriate within-group means and covariances ($k = 8$). Equal sample sizes were chosen for each group, although in real life the normal population would have the largest a priori probability. For one experiment $n_i = 30$, $i = 1, \ldots, 5$, and for the other experiment, $n_i = 100$, $i = 1, \ldots, 5$. Table 5-3 gives the values of the

Table 5-3 Frequency of correct classification

		Linear discriminant			Quadratic discriminant		
		I	II	III	I	II	III
n_i		Actual	Apparent	Independent study	Actual	Apparent	Independent study
30	Mean	.664	.699	.657	.753	.867	.673
	S.D.	.042	.046	.044	.044	.029	.029
100	Mean	.675	.682	.661	.754	.788	.721
	S.D.	.014	.012	.019	.017	.013	.014

correct classification rate of the samples using the model parameters (I), the apparent correct classification rate (II), and the correct classification rate based on classification of an independent sample (III). The last rate (III) is an unbiased estimate of the expected actual error rate. Ten replicates were obtained.

From these results he notes that the quadratic functions perform considerably better than the linear functions on this data. The apparent rate is clearly biased, the mean bias being .035 for the linear functions and .114 for the quadratic functions when $n_i = 30$. When $n_i = 100$ the biases are .007 and .034. The quadratic functions are much more sensitive to bias in the apparent rate, presumably because of the greater number of parameters to be estimated. He points out that the difference between the linear and quadratic functions were not observed in the original ECG data and suggests that the lack of normality in the original data may be the reason for this.

Additional sampling experiments tended to confirm these results. In general, larger sample sizes are needed for good convergence to the optimal correct classification rates for the quadratic function approach than for the linear function approach.

Example

Weather prediction provides a good example of a multiple-group problem. Two examples were given by Miller (*373*). In the first example, he desired to predict airfield ceiling conditions 2 hours in advance. Five ceiling groups were defined as follows:

1. Closed, ceiling < 200 ft.
2. Low instrument, 200 ft $\leq$ ceiling < 500 ft.
3. High instrument, 500 ft $\leq$ ceiling < 1500 ft.

Table 5-4 *Variables used in weather-prediction example*

Variable
Height of lowest cloud layer
Height of second cloud layer
Height of ceiling
Three-hour change in ceiling height
Three-hour change in pressure
East-west wind component
Three-hour change in wind direction
Three-hour change in temperature
Amount of lowest cloud layer
Amount of second cloud layer
Visibility
Three-hour change in visibility
Temperature-dewpoint depression over temperature
North-south wing component
Total cloud cover

4. Low open, 1500 ft $\leq$ ceiling $<$ 5000 ft.

5. High open, 5000 ft $\leq$ ceiling.

He wished to make predictions for McGuire Air Force Base using data from weather stations at McGuire AFB; Newark, New Jersey; Philadelphia, Pennsylvania; Atlantic City, New Jersey; and Lakehurst Naval Air Station, New Jersey. The variables that were available are given in Table 5-4. All fifteen variables were measured at all five stations, giving a total of 75 variables to use in the analysis. The number of observations within each group was as given in Table 5-5.

Miller used the eigenvalue approach to multiple-group discrimination and in selecting variables used a procedure based on giving the largest

Table 5-5

Group	*initial sample*	*Test sample*
(1)	49	35
(2)	84	76
(3)	158	118
(4)	228	124
(5)	1355	573
Total	1874	926

Table 5-6 Predictors selected

Station	Variable
1. Philadelphia	Height of ceiling
2. McGuire AFB	Height of ceiling
3. McGuire AFB	East-west wind component
4. Newark	Height of ceiling
5. Newark	Total cloud cover

increase in tr $W^{-1}B$. The five predictors selected are listed in Table 5-6. Miller points out that there are some variables that are not selected which would seem important on a priori grounds. This is attributed to the fact that there is a good deal of redundancy in the variables. It is of interest to note that the second variable selected is a "persistence" variable. Using it alone, one would predict the ceiling to be what it was 2 h previously. A commonly used criterion in meteorology is to compare the proposed technique with the results predicted by the persistence method.

To compare the performance of the discriminant method with persistence, he assigns all observations and calculates a measure. We present in Table 5-7 the probability of correct classification (not his measure) for various discriminants. The bias of the apparent rate of correct classification is obvious. The performance of the discriminant function is only slightly better than persistence, but it is statistically significant.

The results also indicated the following:

1. Bivariate normality of the eigenvectors is unlikely, as is equality of covariance matrices.

2. When the high open (group 5) conditions hold, there are two clusters of points in the ellipse of concentration. These correspond to clear skies over Philadelphia, McGuire, and Newark, or to high ceilings other than unlimited at Philadelphia, McGuire, and Newark. If "there is disparity among the conditions at Philadelphia, McGuire, and Newark, the observations point is likely to be located in any region of the space. . . ."

Table 5-7 Correct classification probabilities

	Initial sample	*Test sample*
Persistence	.8298	.7473
One eigenvector	.8310	.7570
Two eigenvectors	.8367	.7689
Three eigenvectors	.8421	.7873

Miller's second example used precipitation conditions for the groups. In this he uses seven variables and 25 stations. The general pattern of results is similar to that in the ceiling example except that the apparent rate of correct classification was about .74 for persistence and about .80 for the discriminant function, whereas the rates in the independent test sample were .7285 for persistence and .7330 for four eigenvectors. In this case no persistence variables were used in the discriminant functions.

In these cases it might have been better to use the simpler persistence predictor. It would have been interesting to see how the optimum discriminant based on normality performed. The newer logistic discriminants might also have provided further insight.

Problems

1. Using the trade school data from Chapter 1, calculate the four sets of discriminant coefficients and constants.
2. Show that the rule obtained from $g = 2$ is identical to the one obtained via the linear discriminant function.
3. Find the canonical vectors of $(\mathbf{B} - \gamma\mathbf{W})$ for the technical college data.

Projects

1. Write a program to compute the optimal discriminant for normal variables. Also include the canonical vector approach. Use subroutines supplied by your local computer center when possible, particularly for the calculation of eigenvalues and eigenvectors. Include the calculation for the apparent error rate.

6 *Miscellaneous problems*

This chapter will consider several unconnected topics that merit discussion but not separate chapters. The topics are variable selection, sequential discrimination, logistic models and risk estimation, constrained discrimination, Bayesian methods, and some comments on methodology in simulation studies.

Variable selection

In order to obtain an effective assignment rule (that is, one with a low error rate), the variables must provide information about the two populations which enables assignment to be made. A simple and obvious statistic that might be used is

$$t_i = \frac{\bar{x}_{1i} - \bar{x}_{2i}}{s_i\sqrt{1/n_1 + 1/n_2}} \tag{6-1}$$

where

$$s_i^2 = \frac{1}{n_1 + n_2 - 2}\left[\sum_{k=1}^{n_1} (x_{1ik} - \bar{x}_{1..})^2 + \sum_{k=2}^{n_2} (x_{2ik} - \bar{x}_{2..})^2\right]$$

The t_i's can be used to order the variables in terms of finding the best single discriminator. However, the joint significance level must take into account that k variables are being tested. For example, if a rule is to include all variables such that the overall significance level is α, a simultaneous procedure should be used. The Bonferroni inequalities offer a convenient and simple way of controlling significance levels, particularly when the variables are not independent (*373a*). The rule becomes: Use variable i only if

$$|\, t_i \,| > C_{1-\alpha/2k, n_1+n_2-2} \tag{6-2}$$

where $C_{1-\alpha/2k, n_1+2_n-2}$ is the $(1 - \alpha/2k) \cdot 100$ percentile of the t distribution with $n_1 + n_2 - 2$ df. If k is not too large, this method should be satisfactory. The Mahalanobis distance between Π_1 and Π_2 when only the ith variable is used is

$$D_i^2 = \frac{(\bar{x}_{1i} - \bar{x}_{2i})^2}{s_i^2} = t_i^2\left(\frac{1}{n_1} + \frac{1}{n_2}\right) \tag{6-3}$$

Thus a variable may be highly significant but have a low D^2 because n_1 and n_2 are large. A low D_i^2 implies low discriminating ability, and therefore we may need to use several variables. If the variables are independent, we may simply add the D_i^2 to get for a set of variables, A,

$$D_A^2 = \sum_{i \in A} D_i^2 \tag{6-4}$$

However, it is highly unlikely that the variables will be independent in practice, and thus the more usual form of D^2 is used:

$$D_A^2 = (\bar{\mathbf{x}}_{1A} - \bar{\mathbf{x}}_{2A})'\mathbf{S}_A^{-1}(\bar{\mathbf{x}}_{1A} - \bar{\mathbf{x}}_{2A}) \tag{6-5}$$

where the subscript A denotes only those variables in this set A. Rao (*454*) has given tests for sufficiency of a specified set of variables. These were discussed in Chapter 2.

Because of the dependence of the variables, the best set of two variables may not include the best single variable. In general, the best subset of $r + 1$ variables may not include the best subset of r variables. For this reason, reliance exclusively on t tests is unwise. One could perform tests on the D_A^2 for all possible subsets of the k observed variables, but this involves $2^k - 1$ tests and is not feasible computationally. In addition, the tests will have different degrees of freedom, depending on the number of variables in the model.

Stepwise entry or deletion of variables is used in BMD 07M. This method starts by calculating a statistic for each variable. A variable is included in the equation if (1) it optimizes the statistic, and (2) the statistic exceeds a threshold value. Three possible statistics are available. The first criterion chooses the variable that maximizes the between-group F. The second chooses the variable that minimizes

$$\frac{1}{h_1} \sum_{j \neq l} \frac{1}{1 + D_{jl}/4}$$

where $D_{jl} = (\bar{\mathbf{x}}_j - \bar{\mathbf{x}}_l)'\mathbf{S}^{-1}(\bar{\mathbf{x}}_j - \bar{\mathbf{x}}_l)$ and $h_1 = g(g - 1)/2$. If $g = 2$, this is equivalent to maximizing F. This criterion tends to separate groups that are close together. The third criterion, a generalization of the second, chooses the variable that minimizes

$$\frac{1}{h_1} = \sum_{j \neq l} \frac{\alpha_{jl}}{1 + D_{jl}/4}$$

where α_{jl} are constants specified by the user. After three variables have entered, the variable-deletion feature becomes operative. If, for any variable included in the equation, the statistic fails to exceed a threshold (not necessarily the same as for inclusion), the variable with the poorest value of the statistic is removed from the equation. Thus a discriminant equation is built up. By judicious use of cutoff points it may be possible to avoid "noise" variables in the equation. In practice, the variable-deletion feature is not often used, and the program is allowed to enter all the possible variables. This is not to be recommended, as it frequently leads to a great deal of "noise" in the discriminant function. An empirical rule often followed is: variables may be entered until the significance level becomes greater than .05; no variable should be entered after that point. Some may prefer a more stringent criterion and not enter variables after a level of .01 has been reached.

Various authors have considered the variable-selection problem and particularly broad literature exists on this problem in regression analysis. A few authors have studied the variable selection problem in its discriminant analysis context. Cochran (*111*) notes that an investigator can suggest a list of probably good discriminators and also a list of possibly good discriminators. He comments: "It would be useful to have a sample rule by which such variates can be discarded. . . ." He rejects the use of the t-test criterion because of its dependence on the sample sizes and suggests that any selection rule be based on D_i^2 or D_i. For independent variates the discriminant function is

$$D(\mathbf{x}) = \frac{\mathbf{x}'(\boldsymbol{\mu}_1 - \boldsymbol{\mu}_2)}{\sqrt{(\boldsymbol{\mu}_1 - \boldsymbol{\mu}_2)'(\boldsymbol{\mu}_1 - \boldsymbol{\mu}_2)}} \tag{6-6}$$

where $D(\mathbf{x})$ has been standardized to have unit variance. In this case, if x_1, say, has distance $\delta_1^2 = (\mu_{11} - \mu_{12})^2$ and x_2 has distance $\delta_2^2 = (\mu_{21} - \mu_{22})^2$, then x_1 is equivalent to $m = \delta_1^2/\delta_2^2$ of the second since it is easily seen that the distance between Π_1 and Π_2 is the same whether we use x_1 or m independent variables with between-group distance δ_2^2. If, say, two variables have a distance of $\Sigma\delta_i^2 = 4$ and an additional 10 variates raise the distance to 4.5, it is probably not worthwhile to bother with the additional 10 variates.

The situation is somewhat altered when correlation is present. Assuming that $\delta_i > 0$, if x_1 and x_2 have correlation ρ and between-group distance δ_i^2, the addition of x_2 to the discriminant formed by x_1 increases the squared distance by

$$\frac{(\delta_2 - \rho\delta_1)^2}{1 - \rho^2} = \frac{\delta_1^2(f - \rho)^2}{1 - \rho^2} \tag{6-7}$$

where $f = \delta_2^2/\delta_1^2$. If $\rho = 0$, this is $f^2\delta_1^2$. Thus correlation improves discrimination (over the independent case) if

$$\frac{(f - \rho)^2}{1 - \rho^2} > f^2 \tag{6-8}$$

Any negative correlation is helpful. As $\rho \to -1$ the increase in distance is unbounded. Thus no rule that ignores correlations can be successful in predicting joint discriminating power. For (6-8) to be satisfied when $\rho > 0$, ρ must be greater than $2f/(1 + f^2)$. Fairly large positive correlations have to hold if x_2 is to be helpful. For example, if $f = .1$, ρ must be greater than .2. If $f = 1$, no positive correlation helps, and if $f = 0$, any correlation at all helps. The last case, $f = 0$, occurs when x_2 is a covariate.

Cochran notes that in practice most correlations are positive. He summarizes his recommendations as follows:

1. If most of the correlations among the poor variates and between poor and good variates are positive and not too large, the joint contribution of the poor variates will be less than in the independent case.

2. If the correlations are negative, the joint contribution of the poor variates will be more than in the independent case.

3. Positive correlations have to be quite high if they are to be helpful.

4. Any variable having negative correlation with the good variate will be helpful.

5. If one uses ΣD_i^2 for deciding to include or exclude a set of variables, one is unlikely to go far wrong, because most correlations in practice are small and positive.

Weiner and Dunn (*564*) compared four methods of variable selection: Use of t-statistics (t), calculating the standardized discriminant coefficient (DF), stepwise selection (ST), and random selection (R). They used five data sets having 25 continuous and/or discrete variables. They used the apparent error rate to evaluate the procedures and also classified an independent sample from each population. They chose $n_1 = n_2 = 50$ in all cases, and constructed the "best" discriminant using each of the criteria above. Their results, presented in Table 6-1, make it clear that random selection is not a good way to select variables. There is a serious bias in the apparent error rates which is more pronounced for the 10-variate rule, as might be expected. Use of the t test is a simple means of ranking and seems quite satisfactory for constructing discriminant functions based on a small number of variables. The stepwise procedure is acceptable for building larger discriminants, but caution is also needed in practice.

Carpenter, Strauss, and Bartko (*89a*) used a stepwise discriminant procedure in combination with a t-test procedure to effect a drastic reduction in the number of variates in a medical diagnosis problem. In a study of schizophrenia diagnosis, 1121 patients were used. These were divided into a sample for construction of the discriminant and another for evaluating it. Each sample contained about 405 schizophrenics and 155 nonschizophrenics. A total of 415 variates were measured on each patient. The first step was to select the 150 best variables by means of t tests. Of these, the 69 statistically significant variables were used in a stepwise discriminant analysis, and the 12 most discriminating symptoms were chosen. A simple system was

Table 6-1 *Performance of selection rules: proportions misclassified*

	Apparent error rate selection rule				*Independent-sample selection rule*			
Data set	*t*	*DF*	*ST*	*R*	*t*	*DF*	*ST*	*R*
				10-variate rule				
I	.19	.20	.18	.26	.33	.37	.35	.37
II	.13	.11	.08	.26	.25	.22	.19	.24
III	.35	.35	.32	.42	.46	.45	.44	.51
IV	.13	.11	.10	.25	.21	.22	.21	.37
V	.20	.26	.16	.31	.30	.29	.33	.24
				5-variate rule				
I	.30	.26	.26	.34	.34	.43	.32	.39
II	.18	.14	.16	.31	.27	.24	.26	.38
III	.42	.39	.35	.45	.44	.46	.46	.45
IV	.19	.15	.17	.26	.20	.20	.22	.37
V	.31	.37	.22	.37	.30	.40	.36	.36
				2-variate rule				
I	.31	.33	.33	.45	.34	.44	.37	.45
II	.28	.28	.27	.40	.32	.32	.30	.35
III	.39	.46	.39	.47	.45	.49	.45	.46
IV	.19	.17	.19	.43	.19	.22	.19	.48
V	.28	.35	.28	.37	.30	.35	.35	.41

Source: Weiner and Dunn (*564*).

Table 6-2 *Error rates in schizophrenia diagnosis*

	Initial sample		*Holdout sample*	
Rule[a]	*Schizophrenic*	*Not schizophrenic*	*Schizophrenic*	*Not schizophrenic*
4 or more	.09	.28	.09	.38
5 or more	.20	.13	.19	.22
6 or more	.34	.04	.37	.06
7 or more	.56	.01	.41	.01
8 or more	.77	.00	.80	.00

Source: Adapted from Carpenter et al. (*89a*).

[a] A patient would be diagnosed as schizophrenic if at least the given number of symptoms were present.

desired, so the final rule weighted all symptoms equally and the rule is based on the number of positive signs. Table 6-2 gives the results of the application of the rules to the initial sample and to the holdout sample. Because of the large number of observations there are only small differences between the two samples, although the difference in the not-schizophrenic error rate for the 4-or-more rule is rather large.

Sequential discrimination

Sometimes when the distance between the populations is fairly small, the discriminatory power of the observed variables is insufficient for satisfactory assignment to Π_1 or Π_2. Several sequential approaches have been proposed to avoid this problem. Suppose that we wish to avoid more than ϵ_1 proportion of errors in Π_1 and ϵ_2 in Π_2. If it is possible (and practical) to obtain independent observations on the individual to be assigned, we may use the sequential probability ratio test to assign to Π_1 or Π_2.

The discriminant function is normally distributed with mean $\delta^2/2$ in Π_1, $-\delta^2/2$ in Π_2, and variance δ^2, where $\delta^2 = (\boldsymbol{\mu}_1 - \boldsymbol{\mu}_2)'\boldsymbol{\Sigma}^{-1}(\boldsymbol{\mu}_1 - \boldsymbol{\mu}_2)$. The assignment rule may be described as follows. We perform a sequential likelihood ratio test of the hypothesis $H_0\colon \mathbf{x} \in \Pi_1$ versus $H_1\colon \mathbf{x} \in \Pi_2$. Observe $\mathbf{x}_1$ and calculate

$$A = \frac{1 - \epsilon_2}{\epsilon_1}$$

$$B = \frac{\epsilon_2}{1 - \epsilon_1}$$

$$\lambda_1 = \frac{f_2(D_T(\mathbf{x}_1)\,;\delta^2)}{f_1(D_T(\mathbf{x}_1)\,;\delta^2)} = e^{-D_T(\mathbf{x}_1)} \tag{6-9}$$

Then,

$$\text{if} \quad \lambda_1 \le B, \text{ assign to } \Pi_1$$

$$\text{if} \quad \lambda_1 \ge A, \text{ assign to } \Pi_2$$

Otherwise, take a second observation and compute

$$\lambda_2 = \prod_{i=1}^{2} \frac{f_2(D_T(\mathbf{x}_i)\,;\delta^2)}{f_1(D_T(\mathbf{x}_i)\,;\delta^2)}$$

and compare λ_2 to A and B. Continue taking observations until λ_i is less than B or greater than A. In general we have

$$\lambda_i = e^{-\Sigma D_T(\mathbf{x}_i)} = e^{-nD_T(\bar{\mathbf{x}})} \tag{6-10}$$

More simply the rule is: Assign to Π_1 if after n observations

$$D_T(\bar{\mathbf{x}}) \geq -\frac{1}{n} \ln B \tag{6-11}$$

to Π_2 if

$$D_T(\bar{\mathbf{x}}) \leq -\frac{1}{n} \ln A$$

For example, $\epsilon_1 = \epsilon_2 = .05$,

$$A = 19, \qquad \ln A = 2.944$$

$$B = \tfrac{1}{19}, \qquad \ln B = -2.944$$

This procedure can be used if it is possible to obtain independent replicates of $\mathbf{x}$ on the same individual. If the parameters are not known, one may estimate them by their maximum likelihood estimates and use this rule. This procedure does not involve the a priori probabilities p_1 and p_2 because we are restricting the individual probabilities of misclassification instead.

Mallows (*344*) studied the sequential discrimination problem from a slightly different point of view. Instead of assuming the possibility of replicating the entire vector, $\mathbf{x}$, he considered the case in which the components of $\mathbf{x}$ were obtained sequentially. Mallows's rule is to assign to Π_1 if

$$\ln \left[\frac{f_2(x_1 \cdots x_s)}{f_1(x_1 \cdots x_s)} \right] < \ln A$$

to Π_2 if

$$\ln \left[\frac{f_2(x_1 \cdots x_s)}{f_1(x_1 \cdots x_s)} \right] > \ln B \tag{6-12}$$

and observe x_{s+1} otherwise; A and B are as before. Now if $\mathbf{x}$ is normal,

$$\ln \frac{f_2(x_1 \cdots x_s)}{f_1(x_1 \cdots x_s)} = -[\mathbf{x} - \tfrac{1}{2}(\boldsymbol{\mu}_1 + \boldsymbol{\mu}_2)]'\boldsymbol{\Sigma}^{-1}(\boldsymbol{\mu}_1 - \boldsymbol{\mu}_2) = -D_T(\mathbf{x}) \tag{6-13}$$

where $\mathbf{x}$ has s components. The same constants A and B can be used because the normal distribution can be transformed so that the first s components are independent of the $(s + 1)$st component. He also considered the problem of the best order to take the observations.

Kendall (*289*) suggested a sequential method based on order statistics. His method is to order all variates. Suppose that on x_1 all observations less than a_1 belong to Π_1 and all observations greater than b_1 belong to Π_2. Then a reasonable rule would be

$$\begin{aligned}&\text{assign to } \Pi_1 \text{ if } x_1 < a_1\\ &\text{assign to } \Pi_2 \text{ if } x_1 > b_1\\ &\text{refer to } x_2 \text{ if } a_1 < x_1 < b_1\end{aligned} \tag{6-14}$$

Continue in this manner until all observations are allocated, or until all variables have been used. This method is distribution-free and easy to understand, but it may leave a substantial proportion of observations unassigned. Richards (*462*) suggested examining the distribution of all variables at each step and also using all pairwise joint distributions at each step as a possible remedy. Feldman, Klein, and Honigfeld (*169*) have proposed a similar "successive screening" technique for ordinate data and found its performance satisfactory in comparison with the linear discriminant function.

The usage of sequential discriminants has not been widespread, for several reasons. First, the methods are more complicated and thus less likely to find ready users. Second, in some fields it is important for legal reasons to have a complete record of all variables. Third, the method assumes an infinite number of variables that have discriminating power. This is never the case, and sometimes very few variables contribute anything. A truncated sequential procedure would be useful, but no such methods are available at present.

Logistic discrimination and the estimation of risk

The probability of being a member of Π_i given $\mathbf{x}$ is, according to Bayes theorem,

$$P(\Pi_i \mid \mathbf{x}) = \frac{P(\mathbf{x} \mid \Pi_i)p_i}{P(\mathbf{x} \mid \Pi_1)p_1 + P(\mathbf{x} \mid \Pi_2)p_2} \tag{6-15}$$

if $g = 2$; the probability of belonging to Π_i is for $g > 2$,

$$P(\Pi_i \mid \mathbf{x}) = \frac{P(\mathbf{x} \mid \Pi_i)p_i}{\sum_{i=1}^{g} P(\mathbf{x} \mid \Pi_i)p_i} \tag{6-16}$$

If $P(\mathbf{x} \mid \Pi_i)$ is multivariate normal with mean $\boldsymbol{\mu}_i$ and covariance matrix $\boldsymbol{\Sigma}$, we can write (6-15) as

$$P(\Pi_1 \mid \mathbf{x}) = \exp(\alpha_0 + \boldsymbol{\beta}'\mathbf{x}) P(\Pi_2 \mid \mathbf{x})$$

$$P(\Pi_2 \mid \mathbf{x}) = \frac{1}{1 + \exp(\alpha_0 + \boldsymbol{\beta}'\mathbf{x})} \tag{6-17}$$

and write (6-16) as

$$P(\Pi_i \mid \mathbf{x}) = \exp(\alpha_{0i} + \boldsymbol{\beta}_i'\mathbf{x}) P(\Pi_g \mid \mathbf{x}) \qquad \text{for} \quad i = 1, \ldots, g - 1$$

$$P(\Pi_g \mid \mathbf{x}) = \frac{1}{1 + \sum_{i=1}^{g-1} \exp(\alpha_{0i} + \boldsymbol{\beta}_i'\mathbf{x})} \tag{6-18}$$

Equations (6-17) and (6-18) are called multivariate logistic functions. Model (6-17) was first proposed by Cornfield (*124*) as a feasible alternative to multiway contingency tables when a large number of factors were studied. In this model, if normality holds, we have

$$\boldsymbol{\beta} = \boldsymbol{\Sigma}^{-1}(\boldsymbol{\mu}_1 - \boldsymbol{\mu}_2) \tag{6-19}$$

the usual discriminant coefficients.

The value of α_0 is

$$\alpha_0 = -\tfrac{1}{2}(\boldsymbol{\mu}_1 + \boldsymbol{\mu}_2)'\boldsymbol{\Sigma}^{-1}(\boldsymbol{\mu}_1 - \boldsymbol{\mu}_2) + \ln \frac{p_1}{1 - p_1} \tag{6-20}$$

As they stand, (6-19) and (6-20) require the estimation of $\boldsymbol{\mu}_1$, $\boldsymbol{\mu}_2$, and $\boldsymbol{\Sigma}$ and the inversion of $\boldsymbol{\Sigma}$. This model also yields an estimate of the increase in risk caused by a unit change in one of the variables. It is a little simpler to use the logit of risk:

$$\ln \frac{P(\Pi_1 \mid \mathbf{x})}{1 - P(\Pi_1 \mid \mathbf{x})} = \ln \exp(\alpha_0 + \boldsymbol{\beta}'\mathbf{x}) = \alpha_0 + \boldsymbol{\beta}'\mathbf{x} \tag{6-21}$$

Thus a unit increase in x_i causes a β_i increase in the logit of risk. Since the quantity $P(\Pi_1 \mid \mathbf{x})/[1 - P(\Pi_1 \mid \mathbf{x})]$ is the relative risk of Π_1 given $\mathbf{x}$, a unit increase in x_i causes an e^{β_i} increase in relative risk. If β_i is negative, this is a decrease.

If the distribution of $\mathbf{x}$ is not multivariate normal, one may still formulate (6-15) or (6-16), insert the values of $P(\mathbf{x} \mid \Pi_i)$, and calculate whatever probabilities are desired. In practice this may be quite difficult, since multivariate distributions of dependent variables are not widely known. Cox (*131*), Walker and Duncan (*552*), and Day and Kerridge (*146*) noted that model (6-17) held for a wide variety of situations, which include:

1. Multivariate normal with equal variances in Π_1 and Π_2.
2. Independent Bernoulli variables.
3. Bernoulli variables following a log-linear model with equal second- and higher-order effects.
4. A mixture of situations 1 and 3.

This partially explains why the linear discriminant function is robust for Bernoulli variables. The model is easily extended to include g populations [Cox (*131*), Anderson (*14*)]. Although for the multivariate normal case, the covariance matrix is $\mathbf{\Sigma}$, and the coefficients $\boldsymbol{\beta} = \mathbf{\Sigma}^{-1}(\boldsymbol{\mu}_1 - \boldsymbol{\mu}_2)$ in general, the covariance matrix is not equal to $\mathbf{\Sigma}$ and the coefficients are usually directly estimated by maximum likelihood. Cox (*131*) formulates the following model. Let

$$Y_i = \begin{cases} 1 & \text{if } \mathbf{x}_i \in \Pi_1 \\ 0 & \text{if } \mathbf{x}_i \in \Pi_2 \end{cases}$$

The log likelihood is

$$L(\mathbf{x}_1 \cdots \mathbf{x}_n, \alpha, \boldsymbol{\beta}) = n\alpha_0 + \boldsymbol{\beta}'\mathbf{T} - \sum_{i=1}^{n} \ln[1 + \exp(\alpha_0 + \boldsymbol{\beta}'\mathbf{x}_i)] \quad (6\text{-}22)$$

where

$$\mathbf{T}' = (T_1 \cdots T_k) \qquad \text{and} \qquad T_j = \sum_{i=1}^{n} x_{ij} y_i$$

Thus

$$\frac{\partial L}{\partial \beta_s} = T_s - \sum_{i=1}^{n} \frac{x_{is} \exp(\alpha_0 + \boldsymbol{\beta}'\mathbf{x}_i)}{1 + \exp(\alpha_0 + \boldsymbol{\beta}'\mathbf{x}_i)} \quad (6\text{-}23)$$

These equations must be solved by iterative methods. A good starting value is usually the discriminant coefficients. This method assumes that the sample was obtained from the whole population and that the proportions of each type must be estimated. Anderson (*14*) notes the interesting result that if the populations are sampled separately, the only coefficient that is changed is α_0. Anderson et al. (*17*) discuss this in relation to the diagnosis of Kerato conjunctivitis sicca, an eye disease.

Halperin, Blackwelder, and Verter (*218*) studied the behavior of the multiple logistic function. They considered the two estimates of the coefficients obtained by maximum likelihood and by the discriminant function. The maximum likelihood method is guaranteed to converge to the correct coefficients if the model is correct, although the discriminant coefficient method is computationally simpler. Their results suggested the following when normality does not hold:

1. If $\beta_i = 0$, the maximum likelihood method will tend to estimate β_i by zero, but this is not necessarily so for the discriminant coefficients.

2. If $\beta_i \neq 0$, the discriminant function gives asymptotically biased estimates.

3. Testing significance gives about the same results for both methods.

4. A better fit is usually obtained by the maximum likelihood method.

5. It is theoretically possible for the discriminant function to "give a very poor fit, even if the model holds."

Multivariate analysis was used to assess the risk of coronary heart disease (CHD), as indicated by the data from the Framingham, Massachusetts, study (*535*) and for the Evans County, Georgia, study (*291*). The Framingham study measured the 12-year incidence of CHD, whereas the Evans County study has incidence data for 8 years. The variables used for the analysis are given in Table 6-3. Variables X_8 to X_{20} have the form of cross products in the Evans County study. This allows for nonequality of some of the covariance terms. They are not "interactions" in the sense of experimental design, although they are sometimes thought of in that way.

Table 6-3 *Variables in CHD studies*

Framingham	*Evans County*
1. Age (yr)	1. Age
2. Serum cholesterol (mg/100 ml)	2. Serum cholesterol
3. Systolic blood pressure (mm Hg)	3. Systolic blood pressure
4. Relative weight (100 × actual weight ÷ median for six-height group)	4. Quetelet index (weight/height^2)
5. Hemoglobin (g/100 ml)	5. Diastolic blood pressure
6. Cigarettes per day	6. Smoking history (0 = no; 1 = yes)
(0 = never smoked	7. ECG
1 = less than 1 pack/day	8. $X_4 \cdot X_1/100$
2 = 1 pack/day	9. $X_4 \cdot X_2$
3 = more than 1 pack/day)	10. $X_4 \cdot X_5$
7. ECG (0 = normal; 1 = abnormal)	11. $X_4 \cdot X_3$
	12. $X_3 \cdot X_2$
	13. $X_5 \cdot X_2$
	14. $X_1 \cdot X_5/100$
	15. $X_1 \cdot X_3/100$
	16. $X_1 \cdot X_2/100$
	17. X_1^2
	18. X_2^2
	19. X_5^2
	20. X_3^2

Sources: Kleinbaum et al. (*291*); Truett, Cornfield, and Kannel (*535*).

Table 6-4

	No CHD		*Developed CHD in 12-year period*	
Age	*Men*	*Women*	*Men*	*Women*
All ages	1929	2540	258	129
30–39	749	1824 (30–39 and 40–49 combined)	40	39 (30–39 and 40–49 combined)
40–49	654		88	
50–62	526	716	130	90

Both studies assumed the multiple logistic risk model (6-17) and used the discriminant coefficients to estimate the values of β_0 and $\boldsymbol{\beta}$:

$$\hat{\boldsymbol{\beta}} = \mathbf{S}^{-1}(\bar{\mathbf{x}}_1 - \bar{\mathbf{x}}_2)$$

and

$$\hat{\alpha}_0 = \ln \frac{1 - \hat{p}_1}{\hat{p}_1} - \tfrac{1}{2}(\bar{\mathbf{x}}_1 + \bar{\mathbf{x}}_2)'\mathbf{S}^{-1}(\bar{\mathbf{x}}_1 - \bar{\mathbf{x}}_2) \qquad (6\text{-}24)$$

where $\hat{p}_1 = n_1/n_1 + n_2$. The approximate variance of $\hat{\beta}_i$ is $s^{ii}(1/n_1 + 1/n_2)$, where s^{ii} is the ith diagonal element of $\mathbf{S}^{-1}$. An alternative approach would have been to estimate the coefficients by maximum likelihood, which might have lead to different coefficients and different levels of risk.

The Framingham data were analyzed by sex for all ages and by sex and age for ages 30–39, 40–49, and 50–62. The sample sizes for these groups were as shown in Table 6-4. Race was not a factor in the Framingham study. In the Evans County study, only males were considered, and race (white or black) was used as a stratification factor. The sample sizes were as given in Table 6-5.

By calculating $P(\text{CHD} \mid \mathbf{x})$ for all subjects in the sample and ordering the subjects on the basis of this estimated risk, it is possible to calculate the expected number of cases for each decile of risk and compare the observed number with the expected number of cases. Because the assumption of

Table 6-5

Race	*No CHD*	*Developed CHD in 8-year period*
White	761	71
Black	439	13

Table 6-6 Expected and observed cases in Framingham and Evans County Studies

	Framingham			*Evans County*		
Decile	*Observed*	*Expected*	*Incidence*	*Observed*	*Expected*	*Incidence*
10	82	90.5	37.5	23	23.5	27.5
9	44	47.1	20.1	14	13.0	16.8
8	31	32.6	14.2	12	9.3	14.4
7	33	25.0	15.1	11	7.1	13.2
6	22	19.7	10.1	3	5.5	3.6
5	20	15.0	9.1	3	4.2	3.6
4	13	11.5	5.9	4	3.2	4.8
3	10	8.6	4.6	1	2.4	1.2
2	3	6.0	1.4	0	1.5	0.0
1	0	3.4	0.0	0	0.7	0.0

Sources: Truett, Cornfield, and Kannel (*535*); Kleinbaum et al. (*291*).

normality is possibly invalid, the number of observed cases need not equal the number of expected cases if the discriminant coefficients are used. If the coefficients are estimated by maximum likelihood, the number of cases observed is equal to the number of cases expected. Table 6-6 gives the observed and expected cases for males in Framingham and for white males in Evans County. For the Framingham data, the equation was

$$P(\text{CHD} \mid \mathbf{x}) = [1 + \exp(10.90 - .071x_1 - .011x_2 - .017x_3 - .014x_4 + .084x_5 - .361x_6 - 1.047x_7)]^{-1} \quad (6\text{-}25)$$

The equation for white males in Evans County was derived by a stepwise procedure and

$$P(\text{CHD} \mid \mathbf{x}) = [1 + \exp(5.79 - .029x_{14} - .677x_6 - .770x_7 - .012x_{16})]^{-1} \quad (6\text{-}26)$$

This resulted from using a significance level of $\alpha = .10$. They entered in the order given in (6-26).

In both studies there is an obvious gradient of incidence with increasing risk. The incidence rates in the two studies are similar except for the highest decile. The coefficients represent the amount of increase for the logit of risk caused by a unit increase in the corresponding variable. For example, if there is an ECG abnormality, the logit of risk is increased by 1.05 in the Framingham study and by .77 in the Evans County study.

If the risk were .05 for a subject (logit $P = \ln .05/.95 = -2.944$) with a normal ECG, the logit of risk for a subject with all other variables equal except for an abnormal ECG would be $-2.944 + 1.047 = -1.897$ (in the Framingham study). The risk would be increased to .13. A similar procedure could be followed for the Evans County data, showing an increase in risk from .05 to .10 for the effect of an abnormal ECG.

Truett, Cornfield, and Kannel (*535*) note that their data are well described by the multiple logistic risk function. In spite of the obvious nonnormality of some of the variables, the exponent of the multiple logistic had a fairly symmetrical "normal-looking" distribution, particularly for men.

The Evans County study attempted to determine if the risk factors in blacks and whites were the same. If this were so, then the risk function for whites and blacks would be similar except for the constant term, which would be affected by the observed incidence difference. For black males equations were developed in two ways. First, the stepwise procedure generated the equation

$$P(\text{CHD} \mid \mathbf{x}) = [1 + \exp(5.228 - 2.263x_6 - .011x_8)]^{-1} \qquad (6\text{-}27)$$

which is obviously different from (6-26). With the use of the variables obtained in (6-26) the risk equation is

$$P(\text{CHD} \mid \mathbf{x}) = [1 + \exp(6.025 - .012x_{14} - .150x_6 - 1.073x_7 - .0097x_{16})]^{-1} \qquad (6\text{-}28)$$

None of the coefficients here were significantly different from those in (6-26). Finally, a model was fit using (6-26), but the constant β_0 was calculated using the incidence data for the black males (13 cases, 439 noncases) instead of the values for white males (71 and 761). This gave a fairly good fit. It was felt that the small number of cases in the black male group led to too great uncertainty in the coefficients to be of use. The final conclusion of the paper was that the blacks respond "to the risk factors in a manner similar to the whites but always at a lower level of CHD."

Constrained discrimination

Recall that the optimum assignment rule has the form assign to Π_i if

$$p_i f_i(\mathbf{x}) = \max_j p_j f_j(\mathbf{x}) \qquad j = 1, \ldots, g \qquad (6\text{-}29)$$

when costs of error are assumed equal. On occasion, the error rates may be so large that the rule is of little practical use. For example, if one is trying to detect adverse drug reactions and has p_1 = probability of reaction = .10 and the trinomial distribution shown in Table 6-7, then the

Table 6-7

	A	B	C
Reactors	.1	.1	.8
Nonreactors	.5	.3	.2

optimal rule is to assign all individuals to the nonreactor group. This gives 0 percent error in the nonreactors and 100 percent in the reactors. From the viewpoint of a physician trying to prevent adverse drug reactions, this is an unacceptable procedure. One alternative is to assign costs to the various types of error, but it is often difficult or impossible to do this. A second alternative is to decide the error rates within each group that can be tolerated and choose a rule that satisfies these constraints. Anderson (*13*) showed that the rule which maximizes the probability of correct classification subject to the constraints

$$P_{ij} \leq \alpha_{ij}, \qquad i, j = 1, \ldots, g \quad i \neq j, \tag{6-30}$$

where P_{ij} is the probability that a member of Π_i is assigned to Π_j is constructed as follows. Let

$$L_0(\mathbf{x}) = 0$$

$$L_i(\mathbf{x}) = p_i f_i(\mathbf{x}) - \sum_{j \neq i} \lambda_{ji} f_j(\mathbf{x}) \qquad i = 1, \ldots, g$$

where the λ_{ji} are determined by the α_{ij}. Assign to Π_i if

$$L_i(\mathbf{x}) = \max_j L_j(\mathbf{x})$$

where Π_0 is interpreted to mean: Do not assign $\mathbf{x}$ to any of $\Pi_1, \ldots, \Pi_g$. In general, determination of the λ_{ij}'s is a complicated problem. For the special case $g = 2$ and $f_i(\mathbf{x})$ multivariate normal with mean $\boldsymbol{\mu}_i$ and covariance matrix $\boldsymbol{\Sigma}$, it can be shown that the assignment rule is of one of four types:

1. If the Bayes solution satisfies the constraints $P_1 \leq \alpha_{12}$, $P_2 \leq \alpha_{21}$, use it. That is,

$$P_1 = P\left(D_T(\mathbf{x}) < \ln\left(\frac{p_2}{p_1}\right) \mid \Pi_1\right) \leq \alpha_{12}$$

$$P_2 = P\left(D_T(\mathbf{x}) > \ln\left(\frac{p_2}{p_1}\right) \mid \Pi_2\right) \leq \alpha_{21} \tag{6-31}$$

In this case, all observations will be assigned.

2. There exist constants k_1 and k_2, $k_2 < k_1$, such that

$$\begin{aligned} P(D_T(\mathbf{x}) < k_2 \mid \Pi_1) &= \alpha_{12} \\ P(D_T(\mathbf{x}) > k_1 \mid \Pi_2) &= \alpha_{21} \end{aligned} \tag{6-32}$$

The assignment rule is then:

$$\begin{aligned} &\text{if } D_T(\mathbf{x}) < k_2, \text{ assign to } \Pi_2 \\ &\text{if } D_T(\mathbf{x}) > k_1, \text{ assign to } \Pi_1 \\ &\text{if } k_2 < D_T(\mathbf{x}) < k_1, \text{ do not assign} \end{aligned} \tag{6-33}$$

The overall proportion of unassigned points is

$$p_1 P(k_2 < D_T(\mathbf{x}) < k_1 \mid \Pi_1) + p_2 P(k_2 < D_T(\mathbf{x}) < k_1 \mid \Pi_2) \tag{6-34}$$

3. There exists a constant k_1 such that

$$P(D_T(\mathbf{x}) > k_1 \mid \Pi_2) = \alpha_{21} \qquad \text{and} \qquad P(D_T(\mathbf{x}) < k_1 \mid \Pi_2) < \alpha_{12}$$

The assignment rule is assign to Π_1 if $D_T(\mathbf{x}) > k_1$ and to Π_2 otherwise.

4. There is a constant k_2 such that

$$P(D_T(\mathbf{x}) < k_2 \mid \Pi_1) = \alpha_{12} \qquad \text{and} \qquad P(D_T(\mathbf{x}) > k_2 \mid \Pi_2) = \alpha_{21}$$

The assignment rule is assign to Π_1 if $D_T(\mathbf{x}) > k_2$ and to Π_2 otherwise.

In cases 3 and 4 no points are unassigned. The values of k_1 and k_2 are the solutions of

$$\begin{aligned} \alpha_{21} &= \Phi\left(-\frac{k_1 + \delta^2/2}{\delta}\right) \\ \alpha_{12} &= \Phi\left(\frac{k_2 - \delta^2/2}{\delta}\right) \end{aligned} \tag{6-35}$$

Table 6-8 gives the probability of correct classification for the cases in which rule 2 applies for several values of α_{ij} and δ. Table 6-9 gives the probability of an observation being unassigned for values of α_{ij} and δ.

Bayesian methods

Bayesian methods have been applied to the discriminant problem in two ways. The discriminant function when the parameters are known is

$$D_T(\mathbf{x}) = [\mathbf{x} - \tfrac{1}{2}(\boldsymbol{\mu}_1 + \boldsymbol{\mu}_2)]'\boldsymbol{\Sigma}^{-1}(\boldsymbol{\mu}_1 - \boldsymbol{\mu}_2) \tag{6-36}$$

The "noninformative prior" for $\boldsymbol{\mu}_1$, $\boldsymbol{\mu}_2$, and $\boldsymbol{\Sigma}^{-1}$ is

$$g(\boldsymbol{\mu}_1, \boldsymbol{\mu}_2, \boldsymbol{\Sigma}^{-1}) \propto |\boldsymbol{\Sigma}|^{(k+1)/2} \tag{6-37}$$

Table 6-8 ***Probability of correct classification for rule***[a]

	p	δ .05	1.0	1.5	2.0	2.5	3.0	3.5	4.0	4.5
$\alpha_{12} = .05$,	.10	.126	.259	.442	.639	.804	.912			
$\alpha_{21} = .05$	.50	.126	.259	.442	.639	.804	.912			
$\alpha_{12} = .05$,	.10	.135	.272	.457	.651	.812				
$\alpha_{21} = .10$	.50	.172	.324	.514	.701	.846				
$\alpha_{12} = .10$,	.10	.208	.376	.572	.751	.880				
$\alpha_{21} = .05$	.50	.172	.324	.514	.701	.846				
$\alpha_{12} = .10$,	.10	.217	.389	.586	.764	.888				
$\alpha_{21} = .10$	.50	.217	.389	.586	.764	.888				
$\alpha_{12} = .01$,	.10	.034	.092	.204	.372	.569	.750	.880	.953	.985
$\alpha_{21} = .01$	.50	.034	.092	.204	.372	.569	.750	.880	.953	.985

[a] The probability of correct classification has the form $pA + (1 - p)B$, so it suffices to give the values for two points. Note that if $\alpha_{12} = \alpha_{21}$, $A = B$ and the probability is constant.

Table 6-9 ***Probability of no classification for rule 2***[a]

	p	δ .05	1.0	1.5	2.0	2.5	3.0	3.5	4.0	4.5
$\alpha_{12} = .05$,	.10	.824	.691	.508	.311	.146	.038			
$\alpha_{21} = .05$	.50	.824	.691	.508	.311	.146	.038			
$\alpha_{12} = .05$,	.10	.770	.633	.448	.254	.093				
$\alpha_{21} = .10$	.50	.753	.601	.411	.224	.079				
$\alpha_{12} = .10$,	.10	.737	.569	.373	.194	.065				
$\alpha_{21} = .05$	.50	.753	.601	.411	.224	.079				
$\alpha_{12} = .10$,	.10	.683	.511	.314	.136	.012				
$\alpha_{21} = .10$	.50	.683	.511	.314	.136	.012				
$\alpha_{12} = .01$,	.10	.956	.898	.786	.618	.421	.240	.110	.037	.005
$\alpha_{21} = .01$	.50	.956	.898	.786	.618	.421	.240	.110	.037	.005

[a] The probabilities of no classification have the form $pA + (1 - p)B$, so it suffices to give the values for two points. Note that if $\alpha_{12} = \alpha_{21}$, $A = B$ and the probability is constant.

This leads to the posterior mean of $D_T(\mathbf{x})$,

$$E(D_T(\mathbf{x}) \mid \bar{\mathbf{x}}_1, \bar{\mathbf{x}}_2, \mathbf{S}) = D_S(\mathbf{x}) + \tfrac{1}{2}k\left(\frac{1}{n_2} - \frac{1}{n_1}\right) \tag{6-38}$$

Geisser (*197*) showed that this function minimizes the squared-error loss function. In general, we may consider the case of unequal covariance matrices and means. The optimal rule when parameters are known is

$$Q_T(\mathbf{x}) = \frac{1}{2}\left[\ln\frac{|\mathbf{\Sigma}_1^{-1}|}{|\mathbf{\Sigma}_2^{-1}|} + (\mathbf{x} - \boldsymbol{\mu}_2)'\mathbf{\Sigma}_2^{-1}(\mathbf{x} - \boldsymbol{\mu}_2) - (\mathbf{x} - \boldsymbol{\mu}_1)'\mathbf{\Sigma}_1^{-1}(\mathbf{x} - \boldsymbol{\mu}_1)\right] \tag{6-39}$$

Using the posterior density,

$$g(\boldsymbol{\mu}_i, \mathbf{\Sigma}^{-1}) \propto |\mathbf{\Sigma}_i^{-1}|^{(k+1)/2} \tag{6-40}$$

Enis and Geisser (*165*) showed that the posterior mean of $Q_T(\mathbf{x})$ is

$$Q_S(x) = \frac{1}{2}\left[\ln\frac{|\mathbf{S}_1^{-1}|}{|\mathbf{S}_2^{-1}|} + (\mathbf{x} - \bar{\mathbf{x}}_2)'\mathbf{S}_2^{-1}(\mathbf{x} - \bar{\mathbf{x}}_2) - (\mathbf{x} - \bar{\mathbf{x}}_1)'\mathbf{S}^{-1}(\mathbf{x} - \bar{\mathbf{x}}_1)\right] + h(k, n_1, n_2) \tag{6-41}$$

where $h(k, n_1, n_2)$ is a rather messy function with the property that $h(k, n, n) = 0$. They enumerate a variety of special cases for which $\mathbf{\Sigma}_i$ or $\boldsymbol{\mu}_i$ are known.

The full Bayesian approach uses the posterior distribution of $\mathbf{x}$ for assignment. In this case one has the "predictive density" of $\mathbf{x}$, given the data

$$f(\mathbf{x} \mid \bar{\mathbf{x}}_1, \bar{\mathbf{x}}_2, \mathbf{S}, \Pi_i) = \int f(\mathbf{x} \mid \boldsymbol{\mu}_1, \boldsymbol{\mu}_2, \mathbf{\Sigma}^{-1}, \Pi_i)\, g(\boldsymbol{\mu}_1, \boldsymbol{\mu}_2, \mathbf{\Sigma}^{-1} \mid \bar{\mathbf{x}}_1, \bar{\mathbf{x}}_2, \mathbf{S})\, d\boldsymbol{\mu}_1\, d\boldsymbol{\mu}_2\, d\mathbf{\Sigma}^{-1} \tag{6-42}$$

Assignment is then based on the statistic

$$W = \ln\frac{f(\bar{\mathbf{x}} \mid \bar{\mathbf{x}}_1, \bar{\mathbf{x}}_2, \mathbf{S}, \Pi_1)}{f(\mathbf{x} \mid \bar{\mathbf{x}}_1, \bar{\mathbf{x}}_2, \mathbf{S}, \Pi_2)} \tag{6-43}$$

It was shown by Geisser (*199*) that

$$W = \frac{k}{2}\ln\frac{n_1}{n_2} + \frac{\nu - k + 1}{2}\ln\frac{n_1 + 1}{n_2 + 1} + \frac{\nu + 1}{2}\ln\frac{(n_2 + 1)\nu + n_2(\mathbf{x} - \bar{\mathbf{x}}_2)'\mathbf{S}^{-1}(\mathbf{x} - \bar{\mathbf{x}}_2)}{(n_1 + 1)\nu + n_1(\mathbf{x} - \bar{\mathbf{x}}_1)'\mathbf{S}^{-1}(\mathbf{x} - \bar{\mathbf{x}}_1)} \tag{6-44}$$

where $\nu = n_1 + n_2 - 2$.

Desu and Geisser (*149*) give an example of the use of this method for discriminating monozygotic and dizygotic twins. In this case the means are equal, so all discriminating information is in the differences in the covariance matrices.

The problems involved in the use of Bayesian methods are twofold. First is the selection of the prior distribution. It is unknown what effects improper selection has on the performance of the rule. However, at least in the semi-Bayesian approach [Eq. (6-38)] the results are sufficiently similar to the non-Bayesian approach that there should be no problems on that score. More informative priors might be used if it is felt proper. The second problem is that of complexity of results. The inclusion of prior distributions adds another level of possible confusion for the statistician to explain to his client. The usual problems of nonnormality are here, but it is not known if they affect the Bayesian methods more or less than the classical methods. An advantage that the semi-Bayesian approach gives is that it provides a theoretical justification for using $D_S(\mathbf{x})$ which is lacking in the classical approach.

Some remarks on sampling studies in discriminant analysis

The reader will have noted that the theoretical properties of the sample linear discriminant function tend to be quite complex. For this reason much research in this area is done by Monte Carlo sampling experiments. The usual techniques of Monte Carlo apply here, but as they are sometimes ignored, they will be mentioned on occasion in this discussion.

A general outline of a sampling study might be as follows. A problem is defined. After some thought (ranging from instantly throwing up one's hands to several months of attempted theoretical analysis) it is decided that the problem is theoretically intractable and therefore special cases should be studied by means of sampling. (In a sampling study, it is only possible to deal with special cases, but hopefully one may cover a wide range of interesting situations.) The distributions of random variables must be specified. If the multivariate normal is being used, it is usually possible to transform the distributions so that at least one of the covariance matrices is the identity matrix. In a two-population, equal-covariance problem, the common covariance matrix can usually be taken as the identity. If different covariance matrices are assumed, one may be taken as I and the other as a diagonal matrix D. One mean may be always taken as the origin; other means usually will be on one of the axes. For example, in a two-population problem, $\boldsymbol{\mu}_1 = \mathbf{0}$, $\boldsymbol{\mu}_2' = (\delta, 0, \ldots, 0)$ has the mean of the second population on the x_1 axis. In studies of stepwise discrimination it may be desired to have all components of $\boldsymbol{\mu}_2$ nonzero.

Once the distributions have been specified, it is necessary to have a supply of random numbers with the specified distribution. Usually these are obtained by using the probability integral transform. That is, if $F_i(x)$ is the cumulative distribution of the ith component of $\mathbf{x}$, $F_i(x)$ has a uniform distribution. By obtaining uniform (0, 1) random numbers, one has

$$u = F_i(x_i) \qquad \text{or} \qquad x_i = F_i^{-1}(u) \tag{6-45}$$

If the x_i's are assumed independent, this generates a variable with the appropriate distribution. For example, suppose that $f(x) = \lambda e^{-\lambda x}$; then $F(x) = 1 - e^{-\lambda x}$, and $F^{-1}(u) = -\ln(1-u)/\lambda$ will produce exponential random variables with parameter λ. [It might be noted that if u is uniform (0, 1) so is $1 - u$, so we could also use $F^{-1}(u) = -\ln(u)/\lambda$, which is slightly faster on a computer.] For the normal distribution there are several options available. One may use a series expansion for $F^{-1}(u)$. An alternative method that has been used in the Scientific Subroutine Package for IBM computers is to sum twelve $U(0, 1)$ random variables to obtain a variable with mean 6 and variance 1 and then subtract 6 from it. A third way is by means of the Box–Muller (*63a*) transformation. If u_1 and u_2 are $U(0, 1)$, then

$$\begin{aligned} x_1 &= (-2 \ln u_1)^{1/2} \cos 2\pi u_2 \\ x_2 &= (-2 \ln u_1)^{1/2} \sin 2\pi u_2 \end{aligned} \tag{6-46}$$

are independent normal (0, 1) variates. Of great importance is the generation of the uniform random variables.

A number of methods have been used to estimate error rates. First, the apparent error rate, with all its problems, has been used widely. A second method often encountered is to generate additional data and evaluate the procedure on it. This has been called variously a holdout method, an index sample method, and a separate sample method. In many problems this is the only feasible way to evaluate errors. A third method is based on the fact that if $\mathbf{x}$ is multivariate normal with mean $\boldsymbol{\mu}$ and variance $\boldsymbol{\Sigma}$, then a linear combination $C(\mathbf{x}) = \mathbf{x}'\mathbf{b} + d$ has a univariate normal distribution with mean $\boldsymbol{\mu}'\mathbf{b} + d$ and variance $\mathbf{b}'\boldsymbol{\Sigma}\mathbf{b}$. Thus the probability that a sample discriminant function is less than C_0 for $\mathbf{x} \in \Pi_i$ can be found as follows:

$$\begin{aligned} &P(D_S(\mathbf{x}) < C_0 \mid x \in \Pi_i) \\ &\quad = P[(\mathbf{x} - \tfrac{1}{2}(\bar{\mathbf{x}}_1 + \bar{\mathbf{x}}_2)]'\mathbf{S}^{-1}(\bar{\mathbf{x}}_1 - \bar{\mathbf{x}}_2) < C_0 \mid \Pi_i) \\ &\quad = P\left[\frac{(\mathbf{x} - \boldsymbol{\mu}_i)'\mathbf{S}^{-1}(\bar{\mathbf{x}}_1 - \bar{\mathbf{x}}_2)}{\sqrt{V_D}} < \frac{C_0 - (\boldsymbol{\mu}_i - \frac{1}{2}(\bar{\mathbf{x}}_1 + \bar{\mathbf{x}}_2))'\mathbf{S}^{-1}(\bar{\mathbf{x}}_1 - \bar{\mathbf{x}}_2)}{\sqrt{V_D}}\right] \\ &\quad = \Phi\left[\frac{C_0 - (\boldsymbol{\mu}_1 - \frac{1}{2}(\bar{\mathbf{x}}_1 + \bar{\mathbf{x}}_2))'\mathbf{S}^{-1}(\bar{\mathbf{x}}_1 - \bar{\mathbf{x}}_2)}{\sqrt{V_D}}\right] \end{aligned} \tag{6-47}$$

where $V_D = (\bar{\mathbf{x}}_1 - \bar{\mathbf{x}}_2)'\mathbf{S}^{-1}\mathbf{\Sigma}\mathbf{S}^{-1}(\bar{\mathbf{x}}_1 - \bar{\mathbf{x}}_2)$. This method gives the exact value of the actual probability of misclassification.

Time-dependent data

In intensive-care units it is important to have a current index of the patient's status. This can be done by calculating a function of the variables being monitored at a point in time. But such calculations do not give an indication of changes in a patient's condition over time. Azen and Afifi (*30a*) have suggested the following procedure. At each point in time calculate a discriminant function for the two groups (survival or death). This function itself is a univariate random variable over time. For each patient, calculate the linear regression over time to get a slope and intercept. Find the linear discriminant function to separate the survivors and nonsurvivors using the slopes and intercepts of each patient as the data.

Afifi et al. (*4*) give an example of the use of this method to predict the outcome of comatose patients who had ingested large doses of barbiturate, glutethimide, or meprobamate. Table 6-10 gives the values of means of selected variables at the initial observation and final observation by survival status.

A total of 25 variables were measured for each of 52 patients, 39 women and 13 men. Eighteen of the patients died. The "variables were measured at intervals ranging from every five minutes to every four hours." The variables given in Table 6-9 are the ones that were most important in

Table 6-10 *Initial and final mean values of the most important variables*

	Initial			*Final*		
Variable	*Survivors*	*Deaths*	*P-value*	*Survivors*	*Deaths*	*P-value*
Systolic pressure (SP)	104	86	<.01	122	87	<.001
Mean various pressure (MVP)	6.2	8.5	<.05	5.8	8.6	<.01
pH	7.46	7.37	<.02	7.46	7.31	<.001
Arterial oxygen saturation (SaO_2)	97	92	<.02	98	86	<.001
Arterial blood lactate[a] (LAC)	1.92	3.11	<.05	1.64	4.70	<.001

Source: Afifi et al. (*4*).
[a] Logarithm of variable used.

separating survivors from decedents. The variables were checked for normality, and transformations were applied to those that did not show approximate normality. A logarithmic transformation was satisfactory in all cases.

The table includes only initial and final measurements on each patient, but intermediate values were available so that an index could be calculated at various times if desired. Discriminant functions were calculated using the initial and final values. A stepwise procedure was used. Table 6-11 gives results for several sets of variable combinations.

In the original article the best combinations of three variables were given for initial measurements and the best combination of two variables were given for final measurements. It is clear that substantial improvement in performance is attained by adding variables. It is also clear that the final measurements are better in discriminating as they are made closer to the final outcome point. They therefore suggest using the discriminant based on final observations as their index at intermediate points. This index was

$$Y = .0785\text{SP} + 12.529\text{pH} \tag{6-48}$$

Survival would be predicted if $Y > 100.73$, death otherwise.

This function can be computed at intermediate points in time and a trend computed. They calculated this function every 4 hours. For each

Table 6-11 Performance of discriminant functions

	Probability of correct classification	
Variable(s)	*Initial observations*	*Final observations*
SP	.65	.80
MVP	.62	.66
pH	.65	.76
SaO_2	.64	.74
LAC	.64	.76
SP and MVP		.87
SP and pH		.90
SP and LAC		.87
SP and SaO_2		.92
SP, MVP, and pH	.76	
SP, pH, and LAC	.77	
SP, pH, and SaO_2	.77	

Source: Afifi et al. (*4*).

patient the least-squares estimates of intercept and slope were computed. That is, they fitted the model

$$Y = a + bt \tag{6-49}$$

for each patient. Then they calculated the discriminant function using a and b as the variables and found that

$$Z = .892a + 21.078b \tag{6-50}$$

Survival was predicted if $Z > 90.059$. The probability of correct classification using Z was .83, which is better than that for the function based on the initial measures but not as good as that based on Y. However, Y is strictly usable only in retrospective studies.

The linear function was used because the observations seemed to show a linear trend. Other functions could be used, but there might be some distributional problems involved in calculating the discriminant function in such a case.

Projects

1. Print the t values for each variable in your program. Devise a method to select certain of these based on the t values, and calculate the discriminant function based on these variables only. For example, you might specify that any variable whose to value was greater than t_0 is to be included (t_0 would be an input parameter).
2. Devise a variable selection method for quadratic discrimination and implement it. Note that if only the variances differ, it should be a function of the F statistics, but if both means and variances differ, something more will be needed.
3. Use your linear discriminant program to compute risks as in (6-17).
4. Add the capability to do constrained discrimination to your linear discriminant function program. Print the values of k_1 and k_2 that you calculate from (6-31) and (6-32).

Bibliography

The bibliography that follows is fairly complete up to 1972, although no guarantees can be made. What one individual considers relevant another may not, so there are undoubtedly a few articles that have been omitted because of differences in my taste from the reader's. The major statistical journals searched were:

1. *Annals of Mathematical Statistics*
2. *Journal of the American Statistical Association*
3. *Biometrics*
4. *Journal of the Royal Statistical Society* (Series B and C)
5. *Biometrika*
6. *Technometrics*
7. *Annals of the Institute of Statistical Mathematics, Tokyo*

Other journals that were searched include:

8. *Psychometrika*
9. *IEEE Transactions on Electronic Computers*
10. *IEEE Transactions on Information Theory*

Other bibliographies that have recently appeared include those of Cacoullos (*87*) and Toussaint (*533*). The weakest area of all of these bibliographies is in applications of discriminant analysis. I would be most grateful if readers of this volume would send me references of their work in this area.

Some simple statistics on the bibliography are of interest. Three frequency tables are presented here. The first gives the distribution of the number of publications in which the author is the sole or first name listed. This is a bit unfair to authors who are joint authors and not listed first, but it was the simplest thing to tally by hand. The second table is the frequency distribution of number of publications by year. The third table is the number of publications appearing in selected journals.

Table 1. Number of publications for principal author

1	2	3	4	5	6	7	8	30
266	64	22	9	2	2	1	3	1

Table 2. Number of publications by year of publication

Up to 1930	51–55	56–60	61–65	After 1965
74	42	52	133	278

Table 3. Number of publications by journal

Journal	Number
Annals of Mathematical Statistics	42
Annals of Eugenics	13
Annals of the Institute of Statistical Mathematics, Tokyo	18
Biometrics	30
Biometrika	18
IEEE (*Transactions on Computers* or *Information Theory*)	58
IEEE (other)	16
JASA	18
JRSS (B or C)	20
All medical journals	44
Psychometrika	10
Sankhya	12
Technometrics	13

1. Abend, K., Harley, T. J., and Chandrasekaran, B. (1969). "Comments on the mean accuracy of statistical pattern recognizers," *IEEE Trans. Inform. Theory*, IT-15, pp. 420–423.

1a. Abernathy, J. R., Greenberg, B. G., and Donnelly, J. F. (1966). "Application of discriminant functions in perinatal death and survival," *Amer. J. Obstet. Gynecol.*, 95, pp. 860–867.

2. Adhikari, B. P. (1957). "Analyse discriminante des mesures de probabilité sur un espace abstrait," *Compt. Rend.* pp. 845–846.

3. Afifi, A., and Azen, S. P. (1972). *Statistical Analysis: A Computer Oriented Approach*, New York: Academic Press, Inc.

4. Afifi, A., Sacks, S., Liu, V. Y., Weil, M. H., and Shubin, H. (1971). "Accumulative prognostic index for patients with barbiturate, glutethimide, and meprobamate intoxication," *New Engl. J. Med.*, 285, pp. 1485–1502.

5. Aitkin, M. (1971). "Statistical theory (behavioral science application)," *Ann. Rev. Psychology*, 22, pp. 225–250.

6. Aizerman, M. A., Braverman, E. M., and Rozonder, L. I. (1964). "The probability problem of pattern recognition learning and the method of potential functions," *Autom. Remote Control*, 25, pp. 1175–1190.

7. Albert, A. (1963). "A mathematical theory of pattern recognition," *Ann. Math. Stat.*, 34, pp. 284–299.
8. Albrecht, R., and Werner, W. (1964). "Error analysis of a statistical decision method," *IEEE Trans. Inform. Theory*, IT-10, pp. 34–38.
9. Alexakos, C. E. (1966). "Predictive efficiency of two multivariate statistical techniques in comparison with clinical predictions," *J. Educ. Res.*, 57, pp. 297–306.
10. Allais, D. C. (1964). "Selection of measurements for prediction," Stanford Electron. Lab., Stanford, Calif., Rept. SEL-64-115.
11. Allais, D. C. (1966). "The problem of too many measurements in pattern recognition and prediction," *IEEE Internat. Convention Record*, 14, pt. 2, pp. 124–130.
12. Amari, S. (1967). "A theory of adaptive pattern classifiers," *IEEE Trans. Electron. Computers*, EC-16, pp. 299–307.
13. Anderson, J. A. (1969). "Discrimination between k populations with constraints on the probabilities of misclassification," *J. Roy. Stat. Soc.*, B31, p. 123.
14. Anderson, J. A. (1972). "Separate sample logistic discrimination," *Biometrika*, 59, pp. 19–36.
15. Anderson, J. A., and Boyle, J. A. (1968). "Computer diagnosis—statistical aspects," *Brit. Med. Bull.*, 24, pp. 230–235.
16. Anderson, J. A., and Boyle, J. A. (1968). "A comparison of statistical techniques in the differential diagnosis of nontoxic goitre," *Biometrics*, 24, pp. 103–116.
17. Anderson, J. A., Whaley, K., Williamson, J., and Buchanan, W. W. (1972). "A statistical aid to the diagnosis of keratoconjunctivitis sicca," *Quart. J. Med.*, 41, pp. 175–189.
18. Anderson, T. W. (1951). "Classification by multivariate analysis," *Psychometrika*, 16, pp. 31–50.
19. Anderson, T. W. (1955). "The integral of a symmetric unimodal function over a symmetric convex set and some probability inequalities," *Proc. Amer. Math. Soc.*, 6, pp. 170–176.
20. Anderson, T. W. (1958). *An Introduction to Multivariate Statistical Analysis*, New York: John Wiley & Sons, Inc.
21. Anderson, T. W. (1964). "On Bayes procedures for a problem with choice of observations," *Ann. Math. Stat.*, 35, pp. 1128–1135.
22. Anderson, T. W. (1966). "Some nonparametric multivariate procedures based on statistically equivalent blocks," *Multivariate Anal. Proc. Internatl. Symp.*, Dayton, Ohio, pp. 5–27. New York: Academic Press.
23. Anderson, T. W. (1973). "Asymptotic evaluation of the probabilities of misclassification by linear discriminant functions," in *Discriminant Analysis and Applications*, T. Cacoullos, ed., New York: Academic Press, Inc., pp. 17–35.
24. Anderson, T. W. (1973). "An asymptotic expansion of the distribution of the 'studentized' classification statistic W," *Ann. Stat.* 1, pp. 964–972.
25. Anderson, T. W., Bahadur, R. R. (1962). "Classification into two multivariate normal distributions with different covariance matrices," *Ann. Math. Stat.*, 33, pp. 42–431.

26. Aoyama, H. (1950). "A note on the classification of data," *Ann. Inst. Stat. Math.*, Tokyo, 2, pp. 17–20.
27. Aoyama, H. (1959). "On the evaluation of the risk index of the railroad crossing," *Ann. Inst. Stat. Math.*, Tokyo, 10, pp. 163–180.
28. Armitage, P. (1950). "Sequential analysis with more than two alternative hypotheses and its relations to discriminant function analysis," *J. Roy. Stat. Soc.*, B12. pp. 137–144.
29. Armitage, P., McPherson, C. K., and Copas, J. C. (1969). "Statistical studies of prognosis in advanced breast cancer," *J. Chronic Diseases*, 22, p. 343.
30. Ashton, E. H., Healy, M. J. R., and Lipton, S. (1957). "The descriptive use of discriminant functions in physical anthropology," *J. Roy. Stat. Soc.*, B146, pp. 552–572.

30a. Azen, S. P., and Afifi, A. A. (1972). "Two models for assessing prognosis on the basis of successive observations," *Math. Biosci.*, 14, pp. 169–176.

31. Bahadur, R. R. (1959). "On classification based on responses to *n* dichotomous items," U.S. School of Aviation Medicine, Proj. 60-13. Randolph Field, Texas.
32. Bahadur, R. R. (1961). "A representation of the joint distribution of responses to *n* dichotomous items," in *Studies in Item Analysis and Prediction*, H. Solomon, ed., Stanford, Calif.: Stanford University Press, pp. 158–168.
33. Bahadur, R. R. (1961). "On classification based on response to *n* dichotomous items," in *Studies in Item Analysis and Prediction*, H. Solomon, ed., Stanford, Calif.: Stanford University Press, pp. 177–186.
34. Bailar, B. A. (1972). "The effect of measurement error on discriminant function analysis," Ph.D. dissert., The American University.
35. Bakis, R. (1968). "An experimental study of machine recognition of hand printed numerals," *IEEE Trans. Systems Sci. Cybernetics*, SSC-4, p. 119.
36. Banerjee, K. S., and Marcus, L. F. (1965). "Bounds in a minimax classification procedure," *Biometrika*, 52, pp. 653–654.
37. Barnard, M. M. (1935). "The secular variation of skull characters in four series of Egyptian skulls," *Ann. Eugen.*, 6, pp. 352–371.
38. Baron, D. N., and Fraser, P. M. (1965). "The digital computer in the classification and diagnosis of diseases," *Lancet*, 2, pp. 1066–1069.
39. Bartlett, M. S. (1939). "The standard errors of discriminant function coefficients," *J. Roy. Stat. Soc.*, 6, pp. 169–173.
40. Bartlett, M. S. (1951). "An inverse matrix adjustment arising in discriminant analysis," *Ann. Math. Stat.*, 22, pp. 107–111.
41. Bartlett, M. S. (1951). "The goodness of fit of a single hypothetical discriminant function in the case of several groups," *Ann. Eugen.*, 16(3), pp. 199–214.
42. Bartlett, M. S., and Please, N. W. (1963). "Discrimination in the case of zero mean differences," *Biometrika*, 50, pp. 17–21.
43. Baten, W. D. (1943). "The discriminant function applied to spore measurements," *Mich. Acad. Sci., Arts and Letters*, 29, pp. 3–7.
44. Baten, W. D. (1943). "The use of discriminating functions in comparing judges scores concerning potatoes," *J. Amer. Stat. Assoc.*, 40, pp. 223–228.
45. Baten, W. D., and Beall, G. (1945). "Approximate methods in calculating discriminant functions," *Psychometrika*, 10, pp. 205–218.

46. Baten, W. D., and Dewitt, C. C. (1944). "Use of the discriminant function in the comparison of proximate coal analysis," *J. Ind. Eng. Chem.*, 16, pp. 32–34.
47. Baten, W. D., and Hatcher, H. M. (1944). "Distinguishing method differences by use of discriminant functions," *J. Expt. Educ.*, 12, pp. 184–186.
48. Baten, W. D., Tack, P. I., and Baeder, H. A. (1958). "Testing for differences between methods of preparing fish by use of discriminant function," *Ind. Qual. Control*, 14(7), pp. 7–10.
49. Bauer, R. K. (1954). "Diskrimianzanalyse," *Allgem. Stat. Arch.*, 38, pp. 205–216.
50. Beall, G. (1945). "Approximate methods in calculating discriminant functions," *Psychometrika*, 10, pp. 205–217.
51. Becker, H. C., Nettleton, W. J., Jr., Meyers, P. H., Sweeney, J. W., and Nice, C. M. (1964). "Digital computer determination of a medical diagnostic index directly from chest x-ray images," *IEEE Trans. Biomed. Eng.*, BME-11, pp. 67–72.
52. Behrens, W. W. (1959). "A note on discriminatory analysis," *Biom. Z.*, 1, pp. 3–14 (in German).
53. Bhattacharya, A. (1942). "On discrimination and divergence," *Proc. Indian Sci. Congr.* pt. III, p. 13.
54. Birnbaum, A., and Maxwell, A. M. (1960). "Classification procedures based on Bayes formula," *Appl. Stat.*, 9.
55. Bledsoe, W. W. (1961). "Further results on the N-tuple pattern recognition method," *IRE Trans. Electron. Computers*, EC-10, p. 96.
56. Bledsoe, W. W. (1966). "Some results on multicategory pattern recognition," *J. Assoc. Comput. Mach.*, 3, pp. 304–316.
57. Bol Sev, L. N. (1955). "A nomogram connecting the parameters of a normal distribution with the probabilities for classification into three groups," *Inzh. Sb. Akad. Nauk SSSR*, 21, pp. 212–214.
58. Bol Sev, L. N. (1957). "A nomogram connecting the parameters of a normal distribution with the probabilities for classification into three groups," *Teoriya Veroyatnostei i ee Primeneniya*, 2, pp. 124–126.
59. Bood, D. M., and Baker, C. B. (1958). "Some problems in linear discrimination," *J. Farm Econ.*, 40, pp. 674–683.
60. Bose, R. C., et al., eds. (1970). *Essays in Probability and Statistics*, Roy Memorial Volume, Chapen Hill, N.C.: University of North Carolina Press.
61. Bowker, A. H. (1960). "A representation of Hotelling's T^2 and Anderson's classification statistic W in terms of simple statistics," *Contrib. Prob. Stat.*, pp. 142–149.
62. Bowker, A. H., and Sitgreaves, R. (1961). "An asymptotic expansion for the distribution function of the W-classification statistic," in *Studies in Item Analysis and Prediction*, H. Solomon, ed., Stanford, Calif.: Stanford University Press, pp. 293–310.
62a. Box, G. E. P., and Muller, M. E. (1958). "A note on the generation of random normal deviates," *Ann. Math. Stat.*, 29, pp. 610–611.
63. Boyle, J. A., and Anderson, J. A. (1968). "Computer diagnosis: clinical aspects," *Brit. Med. Bull.*, 24, pp. 224–229.
64. Brier, G. W., Schoot, R. G., and Simmons, V. L. (1940). "The discriminant

function applied to quality rating in sheep," *Proc. Amer. Soc. An. Prod.*, 1, pp. 153–160.

65. Brinegar, C. S. (1963). "Mark Twain and the Quintus Curtiss Snodgrass letters: a statistical test of authorship," *J. Amer. Stat. Assoc.*, 58, pp. 85–96.
66. Brodman, K., et al. (1959). "Interpretation of symptoms with a data processing machine," *A.M.A. Arch. Internal Med.*, 103, pp. 776–782.
67. Brodman, K. (1960). "Diagnostic decisions by machine," *IRE Trans. Med. Electron.*, ME-7, pp. 216–219.
68. Broffitt, J. D. (1969). "Estimating the probability of misclassification based on discriminant function techniques," Ph.D. dissert., Colorado State University.
69. Brogden, H. E. (1955). "Least squares estimates and optimal classification," *Psychometrika*, 20, pp. 249–252.
70. Bromley, D. W. (1971). "The use of discriminant analysis in selecting rural development strategies," *Amer. J. Agr. Econ.* N53, pp. 319–322.
71. Brown, A. M. (1971). "Classification using dichotomous responses," D.Sc. Dissert., University of Pittsburgh.
72. Brown, G. W. (1947). "Discriminant functions," *Ann. Math. Stat.*, 18, p. 514.
73. Brown, G. W. (1950). "Basic principles for construction and application of discriminators," *J. Clin. Psychology*, 6, pp. 58–61.
74. Bryan, J. G. (1951). "The generalized discriminant function: mathematical foundation and computational routine," *Harvard Educ. Rev.*, 21, pp. 90–95.
75. Bryson, M. (1965). "Errors of classification in a binomial population," *J. Amer. Stat. Assoc.*, 60, p. 217.
76. Bulbrook, R. D., Greenwood, F. C., and Hayward, J. L. (1960). "Selection of breast-cancer patients for adrenalectomy or hypophysectomy," *Lancet*, 1, p. 1154.
77. Bulbrook, R. D., Hayward, J. L., Spicer, C. C., and Thomas, B. S. (1962). "A comparison between the urinary steroid excretion of normal women and women with advanced breast cancer," *Lancet*, 2, pp. 1235–1237.
78. Bulbrook, R. D., Hayward, J. L., and Thomas, B. S. (1964). "The relation between the urinary 17-hydroxycorticosteroids and the fate of patients after mastectomy," *Lancet*, 1, p. 945.
79. Bunke, O. (1964). "Uber optimale verfahren der diskriminanzanalyse," *Abhandl. Deut. Akad. Wiss., Kl. Math. Physik Tech.*, pp. 35–41.
80. Bunke, O. (1966). "Nichtparametrische klassifikationsverfahren fur qualitative und quantitative beobachtungen," *Wiss. Z. Humboldt-Univ. Berlin, Math. Natur. Reihe*, 15, pp. 15–18.
81. Bunke, O. (1967). "Stabilitat statistischer entschridungsprobleme und anwendung in der diskriminanzanalyse," *Z. Wahrschein, Theorie und verwandte Gebiete*, 7, pp. 131–146.
82. Burbank, F. (1969). "A computer diagnostic system for the diagnosis of prolonged undifferentiating liver disease," *Amer. J. Med.*, 46, pp. 401–415.
83. Burnaby, T. P. (1966). "Growth invariant discriminant functions and generalized distances," *Biometrics*, 22, p. 96.
84. Burnaby, T. P. (1966). "Distribution-free quadratic discriminant functions in paleontology," *Kansas Geol. Surv. Computer Control.*, pp. 70–77.

85. Burt, C. (1950). "Appendix: on the discrimination between members of two groups," *Brit. J. Psychol. Stat.*, Sect. 3, p. 104.

85a. Cacoullos, T. (1966). "Estimation of a multivariate density," *Ann. Inst. Stat. Math.*, Tokyo, 18, pp. 179–189.

86. Cacoullos, T. (1966). "On a class of admissible partitions," *Ann. Math. Stat.*, 37, pp. 189–195.

87. Cacoullos, T., ed. (1973). *Discriminant Analysis and Applications*, New York: Academic Press, Inc.

88. Calvert, T. (1970). "Nonorthogonal projections for feature extraction in pattern recognition," *IEEE Trans. Computers*, C-19, pp. 447–452.

89. Camp, B. A. (1946). "The effect on a distribution function of small changes in the population," *Ann. Math. Stat.*, 17, pp. 226–231.

89a. Carpenter, W. T., Jr., Strauss, J. S., and Bartko, J. J. (1973). "Flexible system for the diagnosis of schizophrenia: report from the WHO International Pilot Study of Schizophrenia," *Science*, 182, pp. 1275–1278.

90. Casey, R., and Nagy, G. (1966). "Recognition of printed Chinese characters," *IEEE Trans. Electron. Computers*, EC-15, pp. 91–101.

91. Cassie, R. M. (1963). "Multivariate analysis in the interpretation of numerical plankton data," *New Zealand J. Sci.*, 6, pp. 35–59.

92. Chaddha, R. L., and Marcus, L. F. (1968). "An empirical comparison of distance statistics for populations with unequal covariance matrices," *Biometrics*, 24, p. 683.

93. Chan, L. S., and Dunn, O. J. (1972). "The treatment of missing values in discriminant analysis—1. The sampling experiment," *J. Amer. Stat. Assoc.*, 67, pp. 473–477.

94. Chandrasekaran, B. (1971). "Independence of measurements and the mean recognition accuracy," *IEEE Trans. Inform. Theory*, IT-17, pp. 452–456.

95. Chandrasekaran, B., and Jain, A. K. (1972). "Quantization of independent measurement and recognition performance," *IEEE Internat. Symp. Inform. Theory*.

96. Charbonnier, A., Cyffers, B., and Schwartz, D. (1957). "Discrimination entre ictères medicaux et chirurgicaux à partir des résultats de l'analyse electrophorétique des proteïnes du sérum," *Bull. Inst. Internatl. Stat.*, 35, pp. 303–320.

97. Charbonnier, A., Cyffers, B., Schwartz, D., and Vessaereau, A. (1955). "Application of discriminatory analysis to medical diagnostic," *Biometrics*, 11, pp. 553–555.

98. Chatterjee, S., and Barcion, S. (1970). "A non-parametric approach to credit screening," *J. Amer. Stat. Assoc.*, 65, pp. 150–154.

99. Chien, Y-T. (1971). "A sequential decision model for selecting feature subsets in pattern recognition," *IEEE Trans. Computers*, C-20, pp. 282–290.

100. Chow, C. K. (1962). "A recognition method using neighbour dependence," *IRE Trans. Electron. Computers*, EC-11, pp. 683–690.

101. Christensen, C. M. (1953). "Multivariate statistical analysis of difference between pre-professional groups of college students," *J. Exptl. Educ.*, 21, pp. 221–232.

102. Chernoff, H. (1956). "A classification problem," *Tech. Rep.* 33, Appl. Math. Stat. Lab., Stanford University.

103. Chu, J. T., and Cheuh, J. C. (1967). "Error probabilities in decision functions for character recognition," *J. Assoc. Comput. Mach.*, 14, pp. 273–280.
104. Chung, C. S., and Morton, N. E. (1959). "Discrimination of genetic entities in muscular dystrophy," *Amer. J. Human Genet.*, II, pp. 339–359.
105. Clarke, M. R. B. (1971). "Computer developments in research and diagnosis," *Proc. Roy. Soc. Med.*, 64, pp. 819–822.
106. Clemens, B., Linder, J., and Shertyer, B. (1970). "Engineers interest patterns: then and now," *Educ. Psychol. Meas.*, 30, pp. 675–685.
107. Clunies-Ross, C. W., and Riffenburgh, R. H. (1960). "Geometry and linear discrimination," *Biometrika*, 47, pp. 185–189.
108. Cochran, W. G. (1961). "On the performance of the linear discriminant functions," *Bull. Internatl. Stat. Inst.*, 34, pp. 435–447.
109. Cochran, W. G. (1962). "On the performance of the linear discriminant function (report on a discussion of a paper by W. G. Cochran)", *Bull. Internatl. Stat. Inst.*, 35, pp. 157–158.
110. Cochran, W. G. (1964). "Comparison of two methods of handling covariate in discriminatory analysis," *Ann. Inst. Stat. Math.*, Tokyo, 16, pp. 45–53.
111. Cochran, W. G. (1964). "On the performance of the linear discriminant functions," *Technometrics*, 6, pp. 179–190.
112. Cochran, W. G. (1966). "Analyse des classifications d'ordre," *Rev. Stat. Appl.*, 14, pp. 5–17.
113. Cochran, W. G. (1968). "Commentary on estimation of error rates in discriminant analysis," *Technometrics*, 10, p. 204.
114. Cochran, W. G., and Bliss, C. I. (1948). "Discriminant functions with covariance," *Ann. Math. Stat.*, 19, p. 151.
115. Cochran, W. G., and Hopkins, C. E. (1961). "Some classification problems with multivariate qualitative data," *Biometrics*, 17, pp. 10–32.
116. Collen, M. F., Rubin, L., Neyman, J., Dantzig, G. B., Baier, R. M., and Siegelaub, A. B. (1964). "Automated multiphasic screening and diagnosis," *Amer. J. Public Health*, 54, p. 741.
117. Cooley, W. W., and Lohnes, P. R. (1962). *Multivariate Procedures for the Behavioral Sciences*, New York: John Wiley & Sons, Inc.
118. Cooper, P. W. (1962). "The hyperplane in pattern recognition," *Cybernetica*, 5, pp. 215–238.
119. Cooper, P. W. (1962). "The hypersphere in pattern recognition," *Inform. Control*, 5, pp. 324–346.
120. Cooper, P. W. (1963). "Statistical classification with quadratic forms," *Biometrika*, 50, pp. 439–448.
121. Cooper, P. W. (1963). "The multiple category Bayes decision procedure," *IEEE Trans. Electron. Computers*, C-12, p. 18.
122. Cooper, P. W. (1964). "Hyperplanes, hyperspheres and hyperquadratics as decision boundaries," in *Computer and Information Sciences*, (J. T. Tou and R. H. Wilcox, ed.), Washington: Spartan Books.
123. Cooper, P. W. (1965). "Quadratic discriminant functions in pattern recognition," *IEEE Trans. Inform. Theory*, IT-11, pp. 313–315.
124. Cornfield, J. (1962). "Joint dependence of risk of coronary heart disease on

serum cholesterol and systolic blood pressure: a discriminant function analysis," *Proc. Fed. Amer. Soc. Exptl. Biol.*, 21(2), pp. 58–61.

125. Cornfield, J. (1967). "Discriminant functions," *Rev. Internal. Stat. Inst.*, 35. pp. 142–153.

126. Cover, T. M. (1965). "Geometrical and statistical properties of linear inequalities with applications in pattern recognition," *IEEE Trans. Electron. Computers*, EC-14, pp. 326–334.

127. Cover, T. M. (1968). "Rates of covergence of nearest neighbour decision procedures," *Proc. 1st Ann. Hawaii Conf. Systems Theory*, pp. 413–415, Western Periodicals Company.

128. Cover, T. M. (1969). "Learning in pattern recognition," *Methodologies of Pattern Recognition*, S. Watanabe, ed., New York: Academic Press, Inc.

129. Cover, T. M., and Hart, P. E. (1967). "Nearest neighbor pattern classification," *IEEE Trans. Inform. Theory*, IT-13, pp. 21–27.

130. Cox, D. R., and Brandwood, L. (1959). "On a discriminatory problem connected with the works of Plato," *J. Roy. Stat. Soc.*, B21, pp. 195–200.

131. Cox, D. R. (1966). "Some procedures associated with the logistic response curve," in *Research Papers in Statistics: Festschrift for J. Neyman*, pp. 55–71. New York: John Wiley & Sons.

132. Cox, G. M., and Martin, W. P. (1937). "Use of a discriminant function for differentiating soils with different *Axotobacter* populations," *Iowa State Coll. J. Sci.*, 11, pp. 323–332.

133. Cramer, E. M. (1967). "Equivalence of two methods of computing discriminant function coefficients," *Biometrics*, 23, p. 153.

134. Crumb, C. B., Jr., and Ruper, C. E. (1959). "The automatic digital computer as an aid in medical diagnosis," *Proc. Eastern Joint Computer Conference*, pp. 174–179.

135. Cupon, J. (1965). "An asymptotic simultaneous diagonalization procedure for pattern recognition," *Inform. Control*, 8, p. 264.

136. Curnow, R. N. (1970). "A classification problem involving human chromosomes," *Biometrics*, 26, p. 547.

137. Cyffers, B. (1965). "Analyse discriminatoire," *Rev. Stat. Appl.*, 13, pp. 29–46.

138. Cyffers, B. (1965). "Analyse discriminatoire II," *Rev. Stat. Appl.*, 13, pp. 39–65.

139. Dagnelie, P. (1966). "On different methods of numerical classification," *Rev. Stat. Appl.*, 14, pp. 55–75.

140. Das Gupta, S. (1964). "Nonparametric classification rules," *Sankhya*, 26, pp. 25–30.

141. Das Gupta, S. (1965). "Optimum classification rules for classification into two multivariate normal populations," *Ann. Math. Stat.*, 36, pp. 1174–1184.

142. Das Gupta, S. (1968). "Some aspects of discriminant function coefficients," *Sankhya*, A30, pp. 387–406.

142a. Das Gupta, S. (1973). "Theories and methods in classification: a review," in *Discriminant Analysis and Applications*, T. Cacoullos, ed., New York: Academic Press, Inc.

143. Das Gupta, S., and Bhattacharya, P. K. (1964). "Classification between exponential populations," *Sankhya*, A25, pp. 17–24.

143a. Davies, M. G. (1970). "The performance of the linear discriminant function in two variables," *Brit. J. Stat. Psychology*, 23, pp. 165–176.

144. Day, B. B., and Sandomire, M. M. (1942). "Use of the discriminant function for more than two groups," *J. Amer. Stat. Assoc.*, 37, pp. 461–472.

145. Day, N. E. (1969). "Linear and quadratic discrimination in pattern recognition," *IEEE Trans. Inform. Theory*, IT-14, pp. 419–420.

146. Day, N. E., and Kerridge, D. F. (1967). "A general maximum likelihood discriminant," *Biometrics*, 23, pp. 313–323.

147. Defrise-Gussenhoven, E. (1966). "A masculinity–feminity scale based on discriminant functions," *Acta Genet. Stat. Med.*, 16, pp. 198–208.

148. Dempster, A. P. (1964). "Tests for the equality of two covariance matrices in relation to a best linear discriminator analysis," *Ann. Math. Stat.*, 35, pp. 190–199.

149. Desu, M. M., and Geisser, S. (1973). "Methods and applications of equal mean discrimination" in *Discriminant Analysis and Applications*, T. Cacoullos, ed., New York: Academic Press, Inc., pp. 139–159.

150. Dickey, J. M. (1968). "Estimation of disease probabilities conditioned on sympton variables," *Math. Biosci.*, 3, pp. 249–265.

151. Dixon, W. J. (1967). *BMD, Biomedical Computer Programs*, Berkeley, Calif.: University of California Press.

152. Dolby, J. L. (1970). "Statistical aspects of character recognition," *Technometrics*, 12, p. 231.

153. Donchin, E., Callaway, E., and Jones, R. T. (1970). "Audition evoked potential variability in schizophrenia," *Electroencephalog. Clin. Neurophysiol.*, 29, pp. 429–440.

154. Drucker, H. (1969). "Computer optimization of recognition networks," *IEEE Trans. Computers*, C-18, pp. 918–923.

155. Duda, R. O., and Fossen, H. (1966). "Pattern classification by iteratively determined linear and piecewise linear discriminant function," *IEEE Trans. Electron. Computers*, EC-15, pp. 220–223.

156. Dunn, O. J. (1971). "Some expected values for probabilities of correct classification in discriminant analysis," *Technometrics*, 13, p. 345.

157. Dunn, O. J., and Varady, P. D. (1966). "Probabilities of correct classification in discriminant analysis," *Biometrics*, 22, p. 908.

158. Dunsmore, I. R. (1966). "A Bayesian approach to classification," *J. Roy. Stat. Soc.*, B28, p. 568.

159. Elashoff, J. D., Elashoff, R. M., and Goldman, G. E. (1967). "On the choice of variables in classification problems with dichotomous variables," *Biometrika*, 54, p. 668.

160. Ebel, R. L. (1947). "The frequency of errors in the classification of individuals on the basis of fallible test scores," *Educ. Psychol. Meas.*, 7, pp. 725–734.

161. Edwards, C. N. (1969). "Cultural values and role decisions—a study of educated women," *J. Coun. Psyc.*, 16, p. 36. Counseling Psychology.

162. Elfving, G. (1961). "A representation of Hotelling's T^2 and Anderson's classification statistic W in terms of simple statistics," in *Studies in Item Analysis and Prediction*, H. Solomon, ed., Stanford, Calif: Stanford University Press, pp. 285–292.

163. Ellison, B. E. (1962). "A classification problem in which information about alternative distributions is based on samples," *Ann. Math. Stat.*, 33, pp. 213–223.

164. Ellison, B. E. (1965). "Multivariate normal classification with covariance known," *Ann. Math. Stat.*, 36, pp. 1787–1793.

165. Enis, P., and Geisser, S. (1970). "Sample discriminants which minimize posterior squared error," *S. African Stat. J.*, 4, pp. 85–93.

166. Estes, S. E. (1965). "Measurement selection for linear discriminants used in pattern classification," Ph.D. dissert., Stanford University.

167. Eysenck, H. J. (1955). "Psychiatric diagnosis as a psychological and statistical problem," *Psychol. Rept.*, 1, pp. 3–17.

168. Feldman, S., Klein, D. F., and Honigfeld, G. (1969). "A comparison of successive screening and discriminant function techniques in medical taxonomy," *Biometrics*, 25, p. 725.

169. Feldman, S., Klein, D. F., and Honigfeld, G. (1972). "The reliability of a decision tree technique applied to psychiatric diagnosis," *Biometrics*, 28, pp. 831–840.

170. Fisher, O. F. (1949). "Diskriminacni analysa a hodnoceni zkousek schopnosti" ("Discriminatory analysis and the weighting of the results of psychological measurements"), *Stat. Obzor.*, 29, pp. 106–129.

171. Fisher, G. R. (1962). "A discriminant analysis of reporting errors in health interviews," *Appl. Stat.*, 11, pp. 148–163.

172. Fisher, R. A. (1936). "The use of multiple measurement in taxonomic problems," *Ann. Eugen.*, 7, pp. 179–188.

173. Fisher, R. A. (1938). "The statistical utilization of multiple measurements," *Ann. Eugen.*, 8, pp. 376–386.

174. Fisher, R. A. (1940). "The precision of discriminant functions," *Ann. Eugen.*, 10, pp. 422–429.

175. Fix, F., and Hodges, J. L. (1951). "Discriminatory analysis: non-parametric discrimination: small sample performance," U.S. School of Aviation Medicine, Proj. 21-49-004, Rep. 11. Randolph Field, Texas.

176. Fix, E., and Hodges, J. L. (1951). "Non-parametric discrimination: consistency properties," U.S. School of Aviation Medicine, Proj. 21-49-004, Rep. 4. Randolph Field, Texas.

177. Fleiss, J., Spitzer, R. L., Cohen, J., and Endicott, J. (1972). "Three computer diagnosis methods compared," *Arch. Gen. Psychiat.*, 27, pp. 643–649.

178. Foley, D. H. (1972). "Consideration of sample and feature size," *IEEE Trans. Inform. Theory*, IT-18, pp. 618–626.

179. Foley, D. H. (1971). "The probability of error on the design set as a function of the sample size and feature size," Ph.D. dissert., Syracuse University.

180. Francis, I. (1966). "Inference in the classification problem," Ph.D. dissert., Harvard University.

181. Frank, R. E., Massy, W. F., and Morrison, D. G. (1965). "Bias in multiple discriminant analysis," *J. Marketing Res.*, 2, pp. 250–258.

182. Friedman, H. D. (1965). "On the expected error in the probability of misclassification," *Proc. IEEE*, 53, pp. 658–659.

183. Fu, K. S., and Chien, Y. T. (1967). "Sequential recognition using a nonparametric ranking procedure," *IEEE Trans. Inform. Theory*, IT-13, pp. 484–492.

184. Fu, K. S. (1969). "Information processing of remotely sensed agricultural data," *Proc. IEEE*, 57, p. 639.

185. Fu, K. S., and Mendel, J. M., eds. (1971). *Adaptive Learning and Pattern Recognition Systems*, New York: Academic Press.

186. Fu, K. S. (1970). "Recognition," *IEEE Trans. Systems Sci. Cybernetics*, SSC-6, p. 3370.

187. Fukunaga, K. (1972). *Introduction to Statistical Pattern Recognition*, New York: Academic Press, Inc.

188. Fukunaga, K., and Kessel, D. (1972). "Error evaluation and model validation in statistical pattern recognition," School of Electrical Engineering, Purdue University, Tech. Rep. TR-EE 72-23.

189. Fukunaga, K., and Kessell, D. (1971). "Estimation of classification error," *IEEE Trans. Computers*, C-20, pp. 1521–1527.

190. Fukunaga, K., and Krile, T. F. (1969). "Calculation of Bayes recognition error for two multivariate Gaussian distributions," *IEEE Trans. Computers*, C-18, pp. 220–229.

191. Fukunaga, K., and Olson, D. R. (1971). "An algorithm for finding intrinsic dimensionality of data," *IEEE Trans. Computers*, C-20, pp. 176–183.

192. Gaffey, W. R. (1951). "Discriminatory analysis: perfect discrimination as the number of variables increases," U.S. School of Aviation Medicine, Proj. 21-49-004. Randolph Field, Texas.

193. Gardiner, J. (1959). "The use of profiles for differential classification," *Educ. Psychol. Meas.*, 19, pp. 191–205.

194. Garrett, H. E. (1943). "The discriminant function and its rise in psychology," *Psychometrika*, 8, pp. 65–79.

195. Gales, K. (1957). "Discriminant functions of socio-economic class," *Appl. Stat.*, 6, p. 123.

196. Geisser, S. (1964). "Posterior odds for multivariate normal classification," *J. Roy. Stat. Soc.*, B26, pp. 69–76.

197. Geisser, S. (1966). "Predictive discrimination," in *Multivariate Anal. Proc. Internatl. Symp.*, Dayton, Ohio, New York: Academic Press, Inc. pp. 149–163.

198. Geisser, S. (1967). "Estimation associated with linear discriminants," *Ann. Math. Stat.*, 38, pp. 807–817.

199. Geisser, S. (1970). "Discriminatory practices," in *Bayesian Statistics*, R. Collier and F. Meyer, eds., Itasca, Ill.: F. E. Peacock Publishers, Inc., pp. 57–70.

200. Geisser, S., and Desu, M. M. (1968). "Predictive zero mean uniform discrimination," *Biometrika*, 55, pp. 519–524.

201. Gessaman, M. P. (1970). "A consistent nonparametric multivariate density estimator based on statistically equivalent blocks," *Ann. Math. Stat.*, 41, pp. 1344–1346.

202. Gessaman, M. P., and Gessaman, P. H. (1972). "A comparison of some multivariate discrimination procedures," *J. Amer. Stat. Assoc.*, 67, pp. 468–472.

203. Gilbert, E. S. (1968). "On discrimination using qualitative variables," *J. Amer. Stat. Assoc.*, 63, p. 1399.

204. Gilbert, E. S. (1969). "The effect of unequal variance–covariance matrices on Fisher's linear discriminant function," *Biometrics*, 25, pp. 505–516.

205. Giri, N. C. (1964). "On the likelihood ratio test of a normal multivariate testing problem," *Ann. Math. Stat.*, 35, p. 1388.

206. Giri, N. C. (1965). "On the likelihood ratio test of a normal multivariate testing problem. II," *Ann. Math. Stat.*, 36, pp. 1061–1065.

207. Giri, N. C. (1964). "Correction to: On the likelihood ratio test of a normal multivariate testing problem," *Ann. Math. Stat.*, 35, pp. 181–189.

208. Glahn, H. R. (1965). "Objective weather forecasting by statistical methods," *Statistician*, 15, pp. 111–142.

209. Glick, N. (1972). "Sample based classification procedures derived from density estimators," *J. Amer. Stat. Assoc.*, 67, pp. 116–122.

210. Glick, N. (1973). "Sample-based multinomial classification," *Biometrics*, 29, pp. 241–256.

211. Gnanadesikan, R., and Kettenring, J. R. (1972). "Robust estimates, residuals, and outlier detection with multiresponse data," *Biometrics*, 28, pp. 81–124.

212. Goldstein, M. (1972). "*k*-Nearest neighbour classification," *IEEE Trans. Inform. Theory*, IT-18, pp. 627–630.

213. Good, I. J. (1965). "Categorization of classification," *Math. Comput. Sci. Biol. Med.*, pp. 115–128.

214. Gose, E. E. (1969). "Introduction to biological and mechanical pattern recognition," in *Methodologies of Pattern Recognition*, S. Watanabe, ed., New York: Academic Press, Inc.

215. Granlund, G. H. (1972). "Fourier preprocessing for hard print character recognition," *IEEE Trans. Computers*, C-21, pp. 195–201.

216. Greenwood, B. (1969). "Sediment parameters and environment discrimination: an application of multivariate statistics," *Can. J. Earth Sci.*, 6, pp. 1347–1358.

217. Griffin, J. S., Jr., King, J. H., Jr., and Tunis, C. J. (1963). "A pattern identification system using linear decision functions," *IBM Systems J.*, 2, pp. 248–267.

218. Halperin, M., Blackwelder, W. C., and Verter, J. I. (1971). "Estimation of the multivariate logistic risk function: a comparison of the discriminant and maximum likelihood approaches," *J. Chronic Diseases*, 24, pp. 125–158.

219. Han, C-P. (1968). "A note on discrimination in the case of unequal covariance matrices," *Biometrika*, 55, p. 586.

220. Han, C-P. (1969). "Distribution of discriminant function when covariance matrices are proportional," *Ann. Math. Stat.*, 40, pp. 979–985.

221. Hanley, J., Walter, D. O., Rhodes, J. M., and Adey, W. R. (1968). "Chimpanzee performance: computer analysis of electroencephalograms," *Nature*, 220, pp. 879–881.

222. Harley, T. J., Kanal, L. N., and Randall, N. C. (1968). "System considerations for automatic imagery screening," in *Pictorial Pattern Recognition*, G. C. Cheng et al., eds., Washington, D.C.: Thompson Book Co., pp. 15–31.

223. Harper, A. E. (1950). "Discrimination between matched schizophrenia and normals by the Wechsler–Bellevue scale," *J. Consult. Psychology*, 14, pp. 351–357.

224. Harper, A. E. (1950). "Discrimination of the types of schizophrenia by the Wechsler–Bellevue scale," *J. Consult. Psychology*, 14, pp. 290–296.

225. Hart, P. E. (1966). "An asymptotic analysis of the nearest neighbor decision rule," Stanford Tech. Rep. 1828-2, Stanford Electronics Lab., Stanford, Calif.

226. Harter, H. L. (1951). "On the distribution of Wald's classification statistic," *Ann. Math. Stat.*, 22, pp. 58–67.

227. Hayashi, C. (1950). "Approach for quantifying qualitative data from the mathematical-statistical point of view," *Ann. Inst. Stat. Math.*, Tokyo, 2, pp. 35–48.

228. Hayashi, C. (1952). "On the prediction of phenomena from qualitative data and the quantification of qualitative data from the mathematical-statistical point of view," *Ann. Inst. Stat. Math.*, Tokyo, 3, pp. 69–98.

229. Hayashi, C. (1953). "Multidimensional quantification with the applications to analysis of social phenomena," *Ann. Inst. Stat. Math.*, Tokyo, 5, pp. 121–143.

230. Healy, M. J. R. (1965). "Computing a discriminant function from within sample dispersions," *Biometrics*, 21, p. 1011.

231. Hellman, M. E. (1970). "The nearest neighbor classification rule with a reject option," *IEEE Trans. Systems Sci. Cybernetics*, SSC-6, pp. 179–185.

232. Henrichon, E. G., and Fu, K. S. (1969). "A nonparametric partitioning procedure for pattern classification," *IEEE Trans. Computers*, C-18, pp. 614–624.

233. Higgins, G. (1970). "A discriminant analysis of employment in defense and non-defense industries," *J. Amer. Stat. Assoc.*, 65, p. 613.

234. Highleyman, W. H. (1962). "Linear decision functions, with application to pattern recognition," *Proc. IRE*, 50, pp. 1501–1514.

235. Highleyman, W. H. (1962). "The design and analysis of pattern recognition experiments," *Bell System Tech. J.*, 41, pp. 723–744.

236. Hildebrandt, B., Michaelis, J., and Koller, S. (1973). "Die haufigheit der fehlklassifikation bei quadratischen diskriminanzanalyse," *Biom. Z.* 15, pp. 3–12.

237. Hills, M. (1966). "Allocation rules and their error rates," *J. Roy. Stat. Soc.*, B28, p. 1.

238. Hills, M. (1967). "Discrimination and allocation with discrete data," *J. Roy. Stat. Soc.*, C16, pp. 237–250.

239. Hinkley, D. V. (1972). "Time ordered classification," *Biometrika*, 59, pp. 509–523.

240. Ho, Y-C., and Agrawala, A. K. (1968). "On pattern classification algorithm introduction and survey," *Proc. IEEE*, 56, pp. 836–862.

241. Ho, Y. C. (1968). "On pattern classification algorithms—introduction and survey," *IEEE Trans. Autom. Control*, AC13, p. 676.

242. Ho, Y. C. (1969). "On pattern classification algorithms—introduction and survey," *Proc. IEEE*, 56, p. 2101.

243. Hodges, J. L., Jr., et al. (1951). USAF SAM Proj. School of Aviation Medicine 21-49-004, Rept. 1-11. Randolph Field, Texas.

244. Hoel, P. G., and Peterson, R. P. (1949). "A solution to the problem of optimum classification," *Ann. Math. Stat.*, 20, pp. 433–438.

245. Hollingsworth, T. H. (1959). "Using an electronic computer in a problem of medical diagnosis," *J. Roy. Stat. Soc.*, A122, pp. 221–231.

246. Hopkins, C. E. (1964). "Discriminant analysis as an aid to medical diagnosis and taxonomy," *J. Indian Med. Profess.*, p. 5043.

247. Horst, P. (1956). "Multiple classification by the method of least squares," *J. Clin. Psychology*, 12, pp. 3–16.

248. Horst, P. (1956). "Least squares multiple classification for unequal subgroups," *J. Clin. Psychology*, 12, pp. 309–315.

249. Horst, P., and Smith, S. (1950). "The discrimination of two racial samples," *Psychometrika* 15, pp. 271–290.

250. Hudimoto, H. (1956). "On the distribution-free classifications of an individual into one of two groups," *Ann. Inst. Stat. Math.*, Tokyo, 8, pp. 105–112.

251. Hudimoto, H. (1957). "A note on the probability of correct classification when the distributions are not specified," *Ann. Inst. Stat. Math.*, Tokyo, 9, pp. 31–36.

252. Hudimoto, H. (1964). "On a distribution-free two-way classification," *Ann. Inst. Stat. Math.*, Tokyo, 16, pp. 247–253.

253. Hughes, G. F. (1968). "On the mean accuracy of statistical pattern recognizers," *IEEE Trans. Inform. Theory*, IT-14, pp. 55–63.

254. Hughes, G. F., and Lebo, J. A. (1967). "Data reduction using information theoretic techniques," *Rome Air Development Center RADC Rept. TR-67-67*, pp. 45–46.

255. Hughes, G. F. (1969). "Number of pattern classifier design samples per class," *IEEE Trans. Inform. Theory*, pp. 615–618.

256. Hughes, R. E. (1954). "The application of multivariate analysis to (a) problems of classification in ecology; (b) the study of the interrelationships of the plant and environment," *VIII Congr. Internatl. Botanique*, Paris, Sec. 7, 8, pp. 16–18.

257. Hughes, R. E., and Lindley, D. V. (1955). "Application of biometric methods to problems of classification in ecology," *Nature* (London), 175, pp. 806–807.

258. Hughes, W. L., Kalbfleisch, J. M., Brandt, E. N., Jr., and Costilloc, J. P. (1963). "Myocardial infarction prognosis by discriminant analysis," *Arch. Internal Med.*, 111, pp. 338–345.

259. Hussain, A. B. S., Toussaint, G. T., and Donaldson, R. W. (1972). "Results obtained using a simple character recognition procedure on Munson's handprinted data," *IEEE Trans. Computers*, C-21, pp. 201–205.

260. Hussain, A. B. S. (1972). "Sequential decision schemes for statistical pattern recognition problems with dependent and independent hypotheses," Ph.D. dissert., Dept. Elec. Eng., University of British Columbia.

261. Ihm, P. (1965). "Automatic classification in anthropology," in *The Use of Computers in Anthropology*, D. Hymes, ed., Den Haag, The Netherlands: Mouton, pp. 358–376.

262. Isaacson, S. L. (1954). "Problems in classifying populations," in *Statistics and Mathematics in Biology*, pp. 107–117. Ames: Iowa State University Press.

263. Ito, T. (1969). "Note on a class of statistical recognition functions," *IEEE Trans. Computers*, C-18, pp. 76–79.

264. Ivanovitch, B. V. (1954). "Sur la discrimination des ensembles statistiques, *Publ. l'Inst. Stat. l'Univ. Paris*, 3, pp. 207–270.

265. Jackson, E. C. (1968). "Missing values in linear multiple discriminant analysis," *Biometrics*, 24, p. 835.
266. Jackson, R. (1950). "The selection of students for freshman chemistry by means of discriminant functions," *J. Exptl. Educ.*, 18, pp. 209–114.
267. Jenden, D. J., Fairchild, M. D., Mickey, M. R., Silverman, R. W., and Yale, C. (1972). "A multivariate approach to the analysis of drug effects on the electroencephalogram," *Biometrics*, 28, pp. 73–80.
268. Jennison, R. F., Penfold, J. B., and Roberts, J. A. F. (1948). "An application to a laboratory problem of discriminant function analysis involving more than two groups," *Brit. J. Soc. Med.*, 2, pp. 139–148.
269. John, S. (1959). "The distribution of Wald's classification statistics when the dispersion matrix is known," *Sankhya*, 21, pp. 371–376.
270. John, S. (1960). "On some classification problems, I, II," *Sankhya*, 22, pp. 301–308.
271. John, S. (1961). "Corrigenda: On some classification statistics," *Sankhya*, A23, p. 308.
272. John, S. (1961). "Errors in discrimination," *Ann. Math. Stat.*, 32, pp. 1125–1144.
273. John, S. (1963). "On classification by the statistics R and Z," *Ann. Inst. Stat. Math.*, Tokyo, 14, pp. 237–246.
274. John, S. (1964). "Further results on classification by W," *Sankhya*, A26, pp. 39–46.
275. John, S. (1965). "Corrections to: On classification by the statistics R and Z," *Ann. Math. Stat. Math.*, Tokyo, 17, p. 113.
276. John, S. (1966). "On the evaluation of probabilities of convex polyhedra under multivariate normal and t distribution, *J. Roy. Stat. Soc.*, B28, p. 366.
277. Johns, M. V., Jr. (1961). "An empirical Bayes approach to non-parametric two-way classification," in *Studies in Item Analysis and Prediction*, H. Solomon, ed., Stanford, Calif.: Stanford University Press, pp. 221–232.
278. Johnson, M. C. (1955). "Classification by multivariate analysis with objectives of minimizing probability of misclassification," *J. Exptl. Educ.*, pp. 259–264.
279. Johnson, P. (1950). "The quantification of data in discriminant analysis," *J. Amer. Stat. Assoc.*, 45, p. 65.
280. Jutzi, E. (1964). "Anwendung der diskriminanzanalyse in der medizin," *Abhandl. Deut. Akad. Wiss., Kl. Math. Physik Tech.*, pp. 73–74.
281. Kabe, D. G. (1963). "Some results on the distribution of two random matrices used in classification procedures," *Ann. Math. Stat.*, 34, pp. 181–185.
282. Kain, R. Y. (1969). "The mean accuracy of pattern recognizers with many pattern classes," *IEEE Trans. Inform. Theory*, pp. 424–425.
283. Kalmus, H., and Maynard Smith, S. (1965). "The antimode and lines of optimal separation in a genetically determined bimodal distribution, with particular reference to phenylthiocarbamide sensitivity," *Ann. Human Genet.*, 29, pp. 127–138.
284. Kamensky, L. A., and Liu, C. N. (1964). "A theoretical and experimental study of a model for pattern recognition," *Computer Inform. Sci.*, pp. 194–218.

285. Kanal, L. (1962). "Evaluation of a class of pattern recognition networks," in *Biological Prototypes and Synthetic Systems*, New York: Plenum Publishing Corporation, pp. 261–269.

286. Kanal, L., and Nambiar, K. K. (1963). "On the application of discriminant analysis to identification in aerial photography," in *Proc. 7th Natl. Conv. Military Electron.*, Washington, D.C. Spartan Books.

287. Kanal, L., and Chandrasekaran, B. (1968). "On dimensionality and sample size in statistical pattern classification," *Proc. Nat. Electron. Conf.*, 24, pp. 2–7.

288. Keehn, D. G. (1965). "A note on learning for Gaussian properties," *IEEE Trans. Inform. Theory*, pp. 126–132.

288a. Kendall, M. G. (1961). *A Course in Multivariate Analysis*, New York: Hafner Press.

289. Kendall, M. G. (1966). "Discrimination and Classification," in *Multivariate Anal. Proc. Internatl. Symp.*, Dayton, Ohio, New York: Academic Press, Inc., pp. 165–185.

290. Kirsth, R. A. (1964). "Computer interpretation of English text and picture patterns," *IEEE Trans. Electron. Computer*, EC-13, pp. 363–376.

291. Kleinbaum, D. G., Kupper, L. L., Cassel, J. C., and Tyroler, H. A. (1971). "Multivariate analysis of risk of coronary heart disease in Evans County, Georgia," *Arch. Internal Med.*, 128, pp. 943–948.

292. Knussman, R. (1968). "Somato-typology as a biometrical problem," *Biom. Z.*, 10, p. 199.

293. Koontz, W. L. G., and Fukunaga, K. (1972). "Asymptotic analysis of a non-parametric clustering technique," *IEEE Trans. Computers*, C-21, pp. 967–997.

294. Kossack, C. F. (1945). "On the mechanics of classification," *Ann. Math. Stat.*, 16, pp. 95–98.

295. Kossack, C. F. (1949). "Some techniques for simple classification," in *Proc. Berkeley Symp. Math. Stat. Probl., 1945–1946*, Berkeley: University of California Press, pp. 345–352.

296. Kossack, C. F. (1965). "Statistical classification techniques," *IBM Systems J.*, 2, pp. 136–151.

297. Kossack, C. F. (1964). "A handbook of statistical classification techniques," Purdue University, Res. Rep. 601-866.

298. Kowalski, C. J. (1972). "A commentary on the use of multivariate statistical methods in anthropometric research," *Amer. J. Phys. Anthropol.*, 36, p. 119.

299. Kramer, I. R. H., Lucas, R. B., El-Labban, N., and Lister, L. (1970). "Use of discriminant analysis for examining histological features of oral keratoses and lichen planus," *Brit. J. Cancer*, 24, pp. 673–686.

300. Krishnaiah, P. R., ed. (1966). *Multivariate Analysis*, New York: Academic Press, Inc.

301. Krishnaiah, P. R., ed. (1969). *Multivariate Analysis II*, New York: Academic Press, Inc.

302. Krishnaswami, P., and Nath, R. (1968). "Bias in multinomial classification," *J. Amer. Stat. Assoc.*, 63, p. 298.

303. Kronmal, R. A., and Tarter, M. (1968). "The estimation of probability densities and cumulatives by Fourier series methods," *J. Amer. Stat. Assoc.*, 63, pp. 925–952.

304. Kshirsagar, A. M. (1963). "Confidence intervals for discriminant function coefficients," *J. Indian Stat. Assoc.*, 1, pp. 1–7.
305. Kshirsagar, A. M. (1964). "Distribution of the direction and co-linearity factors in discriminant analysis," *Proc. Cambridge Phil. Soc.*, 60, pp. 217–225.
306. Kshirsagar, A. M. (1971). "Goodness of fit of a discriminant function from the vector space of dummy variables," *J. Roy. Stat. Soc.*, B33, p. 111.
307. Kudo, A. (1959). "The classificatory problem viewed as a two-decision problem," pt. I. *Mem. Fac. Sci., Kyushu Univ.*, 13, pp. 96–125 (1959); pt II. *Mem. Fac. Sci., Kyushu Univ.*, 14, pp. 68–73 (1960).
308. Kullback, S. (1965). "On the distribution of two random matrices used in classification procedures," *Ann. Math. Stat.*, 23, pp. 88–102.
309. Kutolin, V. A. (1968). "Statistical petrochemical criteria of foundations for basalts and dobrites, *Dokl. Acad. Nauk. SSSR*, 178, p. 434.
310. Kutolin, V. A. (1969). "A statistical investigation of oxidation of iron in basic rock," *Dokl. Acad. Nauk. SSSR*, 189, p. 173.
311. Lachenbruch, P. A. (1965). "Estimation of error rates in discriminant analysis," Ph.D. dissert., University of California at Los Angeles.
312. Lachenbruch, P. A. (1966). "Discriminant analysis when the initial samples are misclassified," *Technometrics*, 8, p. 657.
313. Lachenbruch, P. A. (1967). "An almost unbiased method of obtaining confidence intervals for the probability of misclassification in discriminant analysis," *Biometrics*, 23, pp. 639–645.
314. Lachenbruch, P. A. (1968). "On expected values of probabilities of misclassification in discriminant analysis, necessary sample size, and a relation with the multiple correlation coefficient," *Biometrics*, 24, p. 823.
314a. Lachenbruch, P. A. (1974). "Discriminant analysis where the initial samples are misclassified. II: Non-random misclassification models," *Technometrics.*
315. Lachenbruch, P. A., and Mickey, M. R. (1968). "Estimation of error rates in discriminant analysis," *Technometrics*, 10, p. 1.
316. Lachenbruch, P. A., Sneeringer, C., and Revo, L. T. (1973). "Robustness of the linear and quadratic discriminant function to certain types of non-normality," *Commun. Stat.*, 1, pp. 39–57,
317. Lachenbruch, P. A. (1973). "Some results on the multiple group discriminant problem," *Discriminant Analysis and Applications*, T. Cacoullos, ed., New York: Academic Press, Inc.
318. Lachenbruch, P. A., and Kupper, L. L. (1973). "Discriminant analysis when one population is a mixture of normals," *Biom. Z.*, 15, pp. 191–197.
319. Ladd, George W. (1966). "Linear probability functions and discriminant functions," *Econometrica*, 34, pp. 873–885.
320. Larsen, L. E., and Walter, D. O. (1970). "On automatic methods of sleep staging by EEG spectra," *Electroencephalog. Clin. Neurophysiol.*, 28, pp. 459–467.
321. Larsen, L. E., and Cornel, J. (1972). "An analytic case study of periparaxysinal events in an implanted semporal lobe epileptic," *Brain Res.* 38, pp. 93–108.
322. Lavelle, G. B. L., Fliman, R. M., Foster, T. D., and Hamilton, M. C. (1970). "An analysis into age changes of human dental arch by a multivariate technique," *Amer. J. Phys. Anthropol.*, 33, pp. 403–411.

323. Lbov, G. S. (1964). "Errors in the classification of patterns for unequal covariance matrices," *Akad. Nauk SSSR. Sibirsk. Inst. Mat. Vyc. Systemy*, 14, pp. 31–38 (in Russian).
324. Ledley, R. S. (1964). "High speed automatic recognition of biomedical pictures," *Science*, 146, pp. 216–223.
325. Ledley, R. S., and Lusted, L. B. (1959). "The use of electronic computers to aid in medical diagnosis," *Proc. IRE*, 47, pp. 1970–1977.
326. Ledley, R. S., and Lusted, L. B. (1959). "Reasoning foundations of medical diagnosis," *Science*, 130, pp. 9–21.
327. Ledley, R. S., and Lusted, L. B. (1963). "Medical diagnosis and modern decision making," in *Proc. Symp. Appl. Math. XIV, Math. Prob. Biol. Sci.*," American Mathematical Society, Providence, R.I., pp. 117–158.
328. Lee, H. C., and Fu, K. S. (1972). "A stochastic syntax analysis procedure and its application to pattern classification," *IEEE Trans. Computers*, C-21, pp. 660–666.
329. Lewis, N. D. C. (1949). "Criteria for early differential diagnosis of psychoneurosis and schizophrenia," *Amer. J. Psychotherapy*, 3, pp. 4–18.
330. Lewis, P. M. (1962). "The characteristic selection problem in recognition systems," *IRE Trans. Inform. Theory*, IT-8, pp. 171–178.
331. Linder, A. (1963). "Trennverfahren bei qualitativen merkmalen, *Metrika*, 6, pp. 76–83.
332. Linhart, H. (1959). "Techniques for discriminant analysis with discrete variables," *Metrika*, 2, pp. 138–149.
333. Linhart, H. (1961). "On the choice of variables in discriminant analysis," *Metrika*, 4, pp. 126–139 (in German).
334. Link, R. F., and Koch, G. S. (1967). "Linear discriminant analysis of multivariate assay and other mineral data," *U.S. Dept. Interior Bur. Mines*, Rept. Invest. N6898.
335. Lohnes, P. R. (1961). "Test space and discriminant space classification models and related significance tests," *Educ. Psychol. Meas.*, 21 pp. 559–574.
335a. Lohnes, P. R., and McIntire, P. H. (1967). "Classification validities of a statewide 10th grade test program," *Personnel Guidance J.*, 45, pp. 561–567.
336. Lu, K. H. (1968). "An information and discriminant analysis of fingerprint patterns pertaining to identification of mongolism and mental retardation," *Amer. J. Human Genet.*, 20, pp. 24–43.
337. Lubin, A. (1960). "Linear and nonlinear discriminatory functions," *Brit. J. Stat. Psychol.*, 3, pp. 90–103.
338. Lubischew, A. A. (1962). "On the use of discriminant functions in taxonomy," *Biometrics*, 18, pp. 455–477.
339. Lusted, L. (1968). *Introduction to Medical Decision Making*, Springfield, Ill.: Charles C Thomas, Publisher.
340. MacNaughton-Smith, P. (1963). "The classification of individuals by the possession of attributes associated with a criterion," *Biometrics*, 17, pp. 364–366.
341. Mahalanobis, P. C. (1936). "On the generalized distance in statistics," *Proc. Natl. Inst. Sci. India*, 2, pp. 49–55.

342. Mahalanobis, P. C. (1940). "The application of statistical methods in physical anthropometry," *Sankhya*, 4, pp. 594–598.
343. Mahalanobis, P. C., Majumdar, D. N., and Rao, C. R. (1949). "Anthropometric survey of the united privinces, 1941: a statistical study," *Sankhya*, 9, pp. 89–324.
344. Mallows, C. L. (1953). "Sequential discrimination," *Sankhya*, 12, p. 321.
345. Marks, S., and Dunn, O. J. (1974). "Discriminant functions when covariance matrices are unequal," *J. Amer. Stat. Assoc.*, 69, pp. 555–559.
346. Marmorsten, J., Weiner, J. M., Hopkins, C. E., and Stern, E. (1966). "Abnormalities in urinary hormone patterns in lung cancer and emphysema," *Cancer*, 19, p. 985.
347. Marmorsten, J., Geller, P. J., and Weiner, J. M. (1969). "Pre-treatment urinary hormone patterns and survival in patients with breast cancer, prostate cancer or lung cancer," *Ann. N.Y. Acad. Sci.*, 164, pp. 483–493.
348. Martin, D. C., and Bradley, R. A. (1972). "Probability models estimation and classification for multivariate dichotomous populations," *Biometrics*, 28, pp. 203–222.
349. Marshall, A. W., and Olkin, I. (1968). "A general approach to some screening and classification problems," *J. Roy. Stat. Soc.*, B30, pp. 407–443.
350. Martin, E. S. (1936). "A study of an Egyptian series of mandibles with special reference to mathematical methods of sexing," *Biometrika*, 28, pp. 149–178.
351. Martin, H. (1960). "Probability of misclassifications. I. Variables," *Qualitatskontrolle und Operations Research*, 5, pp. 109–112 (in German).
352. Massey, W. F. (1965). "Discriminant analysis of audience characteristics," *J. Advertising Res.*, 5, pp. 39–48.
353. Matusita, K. (1956). "Decision rule, based on distance for the classification problem," *Ann. Inst. Stat. Math*, Tokyo, 8, pp. 67–78.
354. Matusita, K. (1966). "A distance and related statistics in multivariate analysis," in *Multivariate Anal. Proc. Internatl. Symp.*, Dayton, Ohio, New York: Academic Press, Inc., pp. 187–200.
355. Matusita, K. (1967). "Classification based on distance in multivariate Gaussian cases," *Proc. 5th Berkeley Symp. Math. Stat. Probl.*, Berkeley: University of California Press.
356. Maung, K. (1941). "Discriminant analysis of Tocher's eye-colour data for Scottish school children," *Ann. Eugen.*, 11, pp. 64–76.
357. McGuire, J. U., and Wirth, W. W. (1958). "The discriminant functions in taxonomic research," in *Proc. 10th Internatl. Congr. Entomol.*, I, pp. 387–393. Paris: Secretariat de l'U.I.S.B.
358. McLachlan, G. J. (1972). "Asymptotic results for discriminant analysis when the initial samples are misclassified," *Technometrics*, 14, pp. 415–422.
359. McLachlan, G. J. (1972). "An asymptotic expansion for the variance of the errors of misclassification of the linear discriminant function," *Austral. J. Stat.*, 14, pp. 68–72.
359a. McLachlan, G. J. (1974). "An asymptotic expansion of the expectation of the estimated error rate in discriminant analysis," *Austral. J. Stat.* (to appear).
359b. McLachlan, G. J. (1974). "An asymptotic unbiased technique for estimating the error rates in discriminant analysis," *Biometrics* 30, 239–249.

359c. McLachlan, G. J. (1974). "Estimation of the errors of misclassification on the criterion of asymptotic mean square error," *Technometrics* 16, 255–260.

360. McQuitty, L. L. (1956). "Agreement analysis: classifying persons by predominant patterns of responses," *Brit. J. Stat. Psychol.*, 9, pp. 5–16.

361. Meisel, W. S. (1968). "Least square methods in abstract pattern recognition," *Inform. Sci.*, 1, p. 43.

362. Meisel, W. S. (1969). "Potential functions in mathematical pattern recognition," *IEEE Trans. Computers*, C-18, pp. 911–918.

362a. Meisel, W. (1972). *Computer Oriented Approaches to Pattern Recognition*, New York: Academic Press, Inc.

363. Mellon, G. B. (1964). "Discriminatory analysis of calcite and silicate cemented phases of the Mountain Park Sandstone," *J. Geol.*, 72, pp. 786–809.

364. Melton, R. S. (1963). "Some remarks on failure to meet assumptions in discriminant analysis," *Psychometrika*, 28, pp. 49–53.

365. Memon, A. Z., and Okamoto, M. (1970). "The classification statistic W^* in covariate discriminant analysis," *Ann. Math. Stat.*, 41, pp. 1491–1499.

366. Merrett, J. D., Wells, R. S., Kerr, C. B., and Barr, A. (1967). "Discriminant function analysis of phenotype variates in ichthyosis," *Amer. J. Human Genet.*, 19, pp. 575–585.

367. Metakides, T. A. (1953). "Calculation and testing of discriminant functions," *Trabajos Estadíst*, 4, pp. 339–368.

368. Meulepas, E. (1967). "Discriminant analysis as a screening test for patients with Turner's syndrome," *Rev. Belge Stat. Rech. Operat.*, 7, pp. 58–64 (in Dutch).

369. Meyer, H. A., and Deming, W. E. (1935). "On the influence of classification on the determination of a measurable characteristic," *J. Amer. Stat. Assoc.*, 30, pp. 671–677.

370. Michaelis, J. (1972). "Zur anwendung der diskriminanzanalyse fur die Medizinische diagnostik," *Habilitationschrift Mainz.*

371. Michaelis, J. (1973). "Simulation experiments with multiple group linear and quadratic discriminant analysis," in *Discriminant Analysis and Applications*, T. Cacoullos, ed., New York: Academic Press, Inc., pp. 225–237.

372. Middleton, G. V. (1962). "Multivariate statistical technique applied to the study of sandstone composition," *Trans. Roy. Soc. Can.*, 56, pp. 119–126.

373. Miller, R. G. (1962). "Statistical prediction by discriminant analysis," *Meterol. Monographs*, 4.

373a. Miller, R. G. (1966). *Simultaneous Statistical Inference*, New York: McGraw-Hill Book Company.

374. Misra, R. K. (1966). "Vectorial analysis for genetic clines in body dimensions in populations of *Drosophila subobscura* Coll and a comparison with those of *D. Robusta* Sturt.," *Biometrics*, 22, pp. 469–487.

375. Mohn, W. S., Jr. (1971). "Two statistical feature evaluation techniques applied to speaker identification," *IEEE Trans. Computers*, C-20, pp. 979–987.

376. Moiola, R. J., and Weiser, D. (1969). "Environmental analysis of ancient sandstone bodies by discriminant analysis," *Bull. Amer. Soc. Petr. Geol.*, 53, p. 733.

376a. Moore, D. H., II (1973). "Combining linear and quadratic discriminants," *Computers Biomed. Res.*, 6, pp. 422–429.

376b. Moore, D. H., II (1973). "Evaluation of five discrimination procedures for binary variables," *J. Amer. Stat. Assoc.*, 68, pp. 399–404.

377. Morris, J. M., Kagan, A., Pattison, D. C., Gardner, M. J., and Raffe, P. A. B. (1966). "Incidence and prediction of ischaemic heart-disease in London busmen," *Lancet*, 2, pp. 553–559.

378. Morrison, D. F. (1967). *Multivariate Statistical Methods*, New York: McGraw-Hill Book Company.

379. Mosteller, F., and Wallace, D. L. (1963). "Inference in an authorship problem," *J. Amer. Stat. Assoc.*, 58, pp. 275–309.

380. Mosteller, F., and Wallace, D. L. (1964). *Inference and Disputed Authorship: The Federalist*, Reading, Mass.: Addison-Wesley Publishing Company, Inc.

381. Motyka, K., and Goszcz, W. (1970). "Application of mathematical methods to evaluation of functional condition of circulatory and respiratory systems," *Acta Physiol. Pol.*, 21, p. 585.

382. Mucciardi, A. N., and Gose, E. E. (1971). "A comparison of seven techniques for choosing subsets of pattern recognition properties," *IEEE Trans. Computers*, C-20, pp. 1023–1031.

383. Munson, J. H., Duda, R. O., and Hart, P. E. (1968). "Experiments with Highleyman's data," *IEEE Trans. Computers*, pp. 399–401.

384. Murthy, V. K. (1965). "Estimation of probability density," *Ann. Math. Stat.*, 36, pp. 1027–1031.

385. Murty, B. R., and Arunachalam, V. (1967). "Computer programmes for some problems in biometrical genetics. I. Use of Mahalanobis' D^2 in classificatory problems," *Indian J. Genet. Plant Breeding*, 27, pp. 60–69.

386. Myers, J. H., and Forgy, E. W. (1963). "The development of numerical credit evaluation systems," *J. Amer. Stat. Assoc.*, 58, pp. 799–806.

387. Nagy, G. (1966). "Self corrective character recognition system," *IEEE Trans. Inform. Theory*, IT-12, p. 215.

388. Nagy, G. (1968). "State of the art in pattern recognition," *Proc. IEEE*, 56, p. 836.

389. Nagy, G. (1968). "Classification algorithm of pattern recognition," *IEEE Trans. Audio*, AU-16, p. 203.

390. Namkoong, G. (1966). "Statistical analysis of introgression," *Biometrics*, 22, pp. 488–502.

391. Nanda, D. N. (1949). "Efficiency of the application of discriminant function in plant selection," *J. Indian Soc. Agr. Stat.*, pp. 8–19.

392. Nanda, D. N. (1949). "The standard errors of discriminant function coefficients in plant-breeding experiments," *J. Roy. Stat. Soc.*, B11, pp. 283–290.

393. Narayan, R. D. (1949). "Some results on discriminant functions," *J. Indian Soc. Agr. Stat.*, 2, pp. 49–59.

394. Nathanson, J. (1971). "Application of multivariate analysis in astronomy," *J. Roy. Stat. Soc.*, C20, p. 239.

395. Noguchi, S., Nagasawa, K., and Oizumi, J. (1969). "The evaluation of the statistical classifier," in *Methodologies of Pattern Recognition*, S. Watanabe, ed., New York: Academic Press, Inc.

396. Ogawa, J. (1960). "A remark on Wald's paper on a statistical problem arising in the classification of an individual into one of two groups," *Collected Papers, 70th Anniv. Nihon Univ. Nat. Sci.*, 3, pp. 12–20. Univ. of Tokyo Press.

397. Okamoto, M. (1961). "Discrimination for variance matrices," *Osaka Math. J.*, 13, pp. 1–39.

398. Okamoto, M. (1963). "An asymptotic expansion for the distribution of the linear discriminant function," *Ann. Math. Stat.*, 34, pp. 1286–1301.

399. Okamoto, M. (1968). "Correction to: An asymptotic expansion for the distribution of the linear discriminant function," *Ann. Math. Stat.*, 39, pp. 1358–1380.

400. Oullett, R. P., and Quadri, S. U. (1968). "Discriminatory power of taxonomic characteristics in separating salmonoid fishes," *System Zool.*, 17, pp. 70–75.

401. Overall, J. E., and Gorham, D. R. (1963). "A pattern probability model for the classification of psychiatric patients," *Behavioral Sciences*, 8, pp. 108–116.

402. Overall, J. E., and Hollister, L. E. (1964). "Computer procedures for psychiatric classification," *J. Amer. Med. Assoc.*, 33, pp. 115–120.

403. Oyama, T., and Tatsuoka, M. (1956). "Prediction of relapse in pulmonary tuberculosis: an application of discriminant analysis," *Amer. Rev. Tuberc. Pulmonary Diseases*," 73, pp. 472–484.

404. Panse, V. G. (1946). "An application of the discriminant function for selection in poultry," *J. Genet.*, 47, pp. 242–248.

405. Palmersheim, J. J. (1970). "Nearest neighbor classification rules: small sample performance and comparison with linear discriminant function and optimal rule," Ph.D. dissert., University of California at Los Angeles.

406. Parzen, E. (1962). "On estimation of a probability density function and mode," *Ann. Math. Stat.*, 33, pp. 1065–1076.

407. Patrick, E. A. (1972). *Fundamentals of Pattern Recognition*, Englewood Cliffs, N.J.: Prentice-Hall, Inc.

408. Patrick, E. A., and Fisher, F. P. (1969). "Non parametric feature selection," *IEEE Trans. Inform. Theory*, IT-15, pp. 577–584.

409. Patrick, E. A., and Fisher, F. P. (1970). "Generalised k nearest neighbour decision rule," *Inform. Control*, 16, pp. 128–152.

410. Pelto, C. R. (1969). "Adaptive nonparametric classification," *Technometrics*, 11, p. 775.

411. Penrose, L. S. (1947). "Some notes on discrimination," *Ann. Eugen.*, 13, pp. 228–237.

412. Peterson, D. W., and Mattson, R. L. (1966). "A method of finding linear discriminant functions for a class of performance criteria," *IEEE Trans. Inform. Theory*, pp. 380–387.

413. Peterson, D. W. (1970). "Some convergence properties of a nearest neighbor decision rule," *IEEE Trans. Inform. Theory*, IT-16, pp. 26–31.

414. Pickrel, E. W. (1958). "Classification theory and techniques," *Educ. Psychol. Meas.*, 18, pp. 37–46.

415. Pipberger, H. W., Klingeman, J. D., and Casma, J. (1968). "Computer evaluation of statistical properties of clinical information in the differential diagnosis of chest pain," *Methods Inform. Med.*, 7, pp. 79–92.

416. Pogue, R. E. (1966). "Some investigation of multivariate discrimination pro-

cedures, with applications to diagnosis clinical electrocardiography," Ph.D. dissert., University of Minnesota.

417. Porebski, O. R. (1966). "On the interrelated nature of the multivariate statistics used in discriminatory analysis," *Brit. J. Math. Stat. Psychology*, 19, pp. 197–214.
418. Porebski, O. R. (1966). "Discriminatory and canonical analysis of technical college data," *Brit. J. Math. Stat. Psychology*, 19, pp. 215–236.
419. Predetti, A. (1960). "On discriminatory analysis," *G. Economisti*, 19, pp. 223–258 (in Italian).
420. Press, S. J. (1968). "Estimating from misclassified data," *J. Amer. Stat. Assoc.*, 63, p. 123.
421. Quenouille, M. H. (1947). "Note on the elimination of insignificant variates in discriminatory analysis," *Ann. Eugen.*, 14, pp. 305–308.
422. Quenouille, M. H. (1948). "A further note on discriminatory analysis," *Ann. Eugen.*, 15, pp. 11–14.
423. Quenouille, M. H. (1968). "The distributions of certain factors occurring in discriminant analysis," *Proc. Cambridge Phil. Soc.*, 64, pp. 731–740.
424. Quesenberry, C. P., and Loftsgaarden, D. O. (1965). "A nonparametric estimate of a multivariate density functions," *Ann. Math. Stat.*, 36, pp. 1049–1051.
425. Quesenberry, C. P., and Gessaman, M. P. (1968). "Nonparametric discrimination using tolerance regions," *Ann. Math. Stat.*, 39, pp. 664–673.
426. Radcliffe, J. (1966). "Factorizations of the residual likelihood criterion in discriminant analysis," *Proc. Cambridge Phil. Soc.*, 62, pp. 743–752.
427. Radcliffe, J. (1967). "A note on an approximate factorization in discriminant analysis," *Biometrika*, 54, p. 665.
428. Radhakrishna, S. (1964). "Discrimination analysis in medicine," *Statistician*, 14, pp. 147–167.
429. Rao, C. R. (1946). "Tests with discriminant functions in multivariate analysis," *Sankhya*, 7, pp. 407–413.
430. Rao, C. R. (1947). "The problem of classification and distance between two populations," *Nature*, 159, p. 30.
431. Rao, C. R. (1947). "A statistical criterion to determine the group to which an individual belongs," *Nature*, 160, pp. 835–836.
432. Rao, C. R. (1948). "Tests of significance in multivariate analysis," *Biometrika*, 35, pp. 58–79.
433. Rao, C. R. (1948). "The utilization of multiple measurements in problems of biological classification," *J. Roy. Stat. Soc.*, B10, pp. 159–193.
434. Rao, C. R. (1949). "On the distance between two populations," *Sankhya*, 9, pp. 246–248.
435. Rao, C. R. (1949). "On some problems arising out of discrimination with multiple characters," *Sankhya*, 9, p. 343.
436. Rao, C. R. (1950). "A note on the distribution of $D^2_{p+q} - D_p{}^2$ and some computational aspects of D^2 statistic and discriminant function," *Sankhya*, 10, pp. 257–268.
437. Rao, C. R. (1950). "Statistical inference applied to classificatory problems," *Sankhya*, 10, pp. 229–256.

438. Rao, C. R. (1951). "Statistical inference applied to classificatory problems. II. The problem of selecting individuals for various duties in a specified ratio," *Sankhya*, 11, pp. 107–116.

439. Rao, C. R. (1951). "Statistical inference applied to classificatory problems. III. The discriminant function approach in the classification of time series," *Sankhya*, 11, pp. 257–272.

440. Rao, C. R. (1952). *Advanced Statistical Methods in Biometric Research*, New York: John Wiley & Sons, Inc.

441. Rao, C. R. (1953). "Statistical inference applied to classificatory problems. Discriminant function for genetic differentiation and selection," *Sankhya*, 12, pp. 229–246.

442. Rao, C. R. (1954). "A general theory of discrimination when the information about alternative population distributions is based on samples," *Ann. Math. Stat.*, 25, pp. 651–670.

443. Rao, C. R. (1954). "On the use and interpretation of distance functions in statistics," *Bull. Internatl. Stat. Inst.*, 34, pp. 90–100.

444. Rao, C. R. (1960). "Multivariate analysis: an indispensible statistical aid in applied research," *Sankhya*, 22, pp. 317–338.

445. Rao, C. R. (1961). "Some observations on multivariate statistical methods in anthropological research," *Bull. Internatl. Stat. Inst.*, 37, pp. 99–109.

446. Rao, C. R. (1962). "Use of discriminant and allied forms in multivariate analysis," *Sankhya*, 24, pp. 149–154.

447. Rao, C. R. (1962). "Some observations in anthropometric surveys," *Indian Anthropol. Essays Mem.*, pp. 135–149.

448. Rao, C. R. (1964). "The use and interpretation of principal components analysis in applied research," *Sankhya*, 26, pp. 329–358.

449. Rao, C. R. (1965). *Linear Statistical Inference and Its Applications*, New York: John Wiley & Sons, Inc.

450. Rao, C. R. (1966). "Discrimination among groups and assigning new individuals," *The Role of Methodology of Classification in Psychiatry and Psychopathology*, pp. 229–240.

451. Rao, C. R. (1966). "Covariance adjustment and related problems in multivariate analysis," *Multivariate Anal. Proc. Internatl. Symp.*, pp. 87–103. New York: Academic Press.

452. Rao, C. R. (1966). "Discriminant function between composite hypotheses and related problems," *Biometrika*, 53, p. 339.

453. Rao, C. R. (1969). "Recent advances in discriminatory analysis," *J. Indian Soc. Agr. Stat.*, 21, pp. 3–15.

454. Rao, C. R. (1970). "Inference on discriminant function coefficients," *Essays in Probability and Statistics*, R. C. Bose et al., eds., Chapel Hill: University of North Carolina and Statistical Publishing Society, pp. 587–602.

455. Rao, C. R. (1972). "Recent trends of research work in multivariate analysis," *Biometrics*, 28, pp. 3–22.

456. Rao, C. R., and Shaw, D. C. (1948). "On a formula for the prediction of cranial capacity," *Biometrics*, 4, pp. 247–253.

457. Rao, C. R., and Slater, P. (1949). "Multivariate analysis applied to differences between neurotic groups," *Brit. J. Psychol. Stat.*, Sect. 2, pp. 17–29.

458. Rao, C. R., and Varadriajan, V. S. (1963). "Discrimination of Gaussian processes," *Sankhya*, 25, pp. 303–330.
459. Rao, M. M. (1963). "Discriminant analysis," *Ann. Inst. Stat. Math.*, Tokyo, 15, pp. 11–24.
460. Revo, L. T. (1970). "On classifying with certain types of ordered qualitative variates: an evaluation of several procedures," *North Carolina Inst. Stat. Mimeo Ser.*, 708.
461. Reyment, R. A. (1966). "Homogeneity of covariance matrices in relation to generalized distances and discriminant functions," *Kansas Geol. Surv. Computer Control*, 7, pp. 5–9.
462. Richards, L. E. (1972). "Refinement and extension of distribution free discriminant analysis," *Appl. Stat.*, 21, pp. 174–176.
463. Riffenburgh, R. H., and Clunies-Ross, C. W. (1960). "Linear discriminant analysis," *Pacific Sci.*, 14, pp. 251–256.
464. Rightmire, G. P. (1970). "Iron age skulls from Southern Africa reassessed by multiple discriminant analysis," *Amer. J. Phys. Anthropol.*, 33, pp. 147–168.
465. Rightmire, G. P. (1970). "Bushman, Hottentot and South Africa negro crania studied by distance and discrimination," *Amer. J. Phys. Anthropol.*, 33, pp. 169–196.
466. Romanovksy, V. I. (1925). "On the statistical criteria that a given specimen belongs to one of allied species," *Turpest. Mautsch. Obstsh pri Sredne-Asiatsk. Gosudarstuv. Univers.* 2, pp. 173–184 (in Russian).
467. Rose, M. J. (1964). "Classification of a set of elements," *Computer J.*, 7, pp. 208–211.
468. Rouvier, R. (1966). "L'analyse en composantes principales nonutilisation en génétique et ses rapports sur l'analyse discriminante," *Biometrics*, 22, p. 343.
469. Rulon, P. J. (1951). "Distinctions between discriminant and regression analysis and a geometric interpretation of the discriminant function," *Harvard Educ. Rev.*, 21, pp. 80–89.
470. Rutowitz, D. (1967). "Pattern recognition," *J. Roy. Stat. Soc.*, A129, pp. 504–530.
471. Sakino, S., and Kano, G. (1954). "On the forecasting of prognosis in pediatrics by a quantifying methods," *Ann. Inst. Stat. Math.*, Tokyo, 6, pp. 173–180.
472. Salvemini, T. (1961). "On the discrimination between two simple and multiple statistical variates," *Statistica*, 21, pp. 121–144 (in Italian).
473. Samuel, E. (1963). "Note on a sequential classification problem," *Ann. Math. Stat.*, 34, pp. 1095–1097.
474. Sammon, J. W. (1970). "Interactive pattern analysis and classification," *IEEE Trans. Computers*, C-19, pp. 594–616.
475. Sammon, J. W. (1970). "An optimal discriminant plane," *IEEE Trans. Computers*, C-19, pp. 826–829.
476. Sammon, J. W., Foley, D. H., and Proctor, A. (1970). "Considerations of dimensionality vs. sample size," *Proc. Symp. Adaptive Processes*, Austin, Tex.
477. Sardar, P. K., Bidwell, O. W., and Marcus, L. F. (1966). "Selection of characteristics for numerical classification of soils," *Soil Sci. Amer. Proc.*, 30, pp. 269–272.

478. Sarndal, C. E. (1967). "On deciding cases of disputed authorship," *J. Roy. Stat. Soc.*, C16, pp. 251–268.

479. Satz, P. (1966). "Specific and nonspecific effects of brain lesions in man," *J. Abnormal Psychology*, 71, pp. 65–70.

480. Saxena, A. K. (1967). "A note on classification," *Ann. Math. Stat.*, 38, pp. 1592–1593.

481. Schmid, J., Jr. (1951). "A comparison of two procedures for calculating discriminant function coefficients," *Psychometrika*, 15, pp. 431–434.

482. Scott, A. J., and Symons, M. J. (1971). "Clustering methods based on likelihood ratio criteria," *Biometrics*, 27, pp. 387–397.

483. Sebestyen, G. (1961). "Recognition of membership in classes," *IRE Trans. Inform. Theory*, pp. 44–50.

484. Sebestyen, G. (1962). *Decision-Making Processes in Pattern Recognition*, New York: Macmillan Publishing Co., Inc.

485. Sebestyen, G., and Edie, J. (1966). "An algorithm for non-parametric pattern recognition," *IEEE Trans. Electron. Computers*, EC-15, pp. 908–915.

486. Sedransk, N. (1969). "Contributions to discriminant analysis," Ph.D. dissert., Iowa State University.

487. Sedransk, N., and Okamoto, M. (1971). "Estimation of the probabilities of misclassification for a linear discriminant function in the univariate normal case," *Ann. Inst. Stat. Math.*, Tokyo, 23(3), pp. 419–435.

488. Shubin, H., Afifi, A., Rand, W. M., and Weil, M. H. (1968). "Objective index of haemodynamic status for quantitation of severity and prognosis of shock complicating myocardial infarction," *Cardiovascular Res.*, 2, p. 329.

489. Sitgreaves, R. (1952). "On the distribution of two random matrices used in classification procedures," *Ann. Math. Stat.*, 23, pp. 263–270.

490. Sitgreaves, R. (1961). "Some results on the distribution of the W-classification statistic," in *Studies in Item Analysis and Prediction*, H. Solomon, ed., Stanford, Calif.: Stanford University Press, pp. 241–251.

491. Smith, C. A. B. (1947). "Some examples of discrimination," *Ann. Eugen.*, 18, pp. 272–283.

492. Smith, F. W. (1968). "Pattern classifier design by linear programming," *IEEE Trans. Computers*, C-17, pp. 367–372.

493. Smith, F. W. (1969). "Design of multicategory pattern classifiers with two-category classifier design procedures," *IEEE Trans. Computers*, C-18, pp. 548–551.

494. Smith, H. F. (1936). "A discriminant function for plant selection," *Ann. Eugen.*, 7, pp. 240–250.

495. Smith, J. E. K., and Klem, L. (1961). "Vowel recognition using a multiple discriminant function," *J. Accoust. Soc. Amer.*, 33, p. 358.

496. Smith, M. S. (1954). "Discrimination between electroencephalograph recordings of normal females and normal males," *Ann. Eugen.*, 18, pp. 344–350.

497. Solomon, H. (1956). "Probability and statistics in psychometric research: item analysis and classification techniques," in *Proc. 3rd Berkeley Symp. Math. Stat. Probl.*, Berkeley: University of California Press, 5, pp. 169–184.

498. Solomon, H. (1960). "Classification procedures based on dichotomous response vectors," *Contributions to Probability and Statistics*, Chapel Hill, University of North Carolina Press, pp. 414–423.

499. Solomon, H., ed. (1961). *Studies in Item Analysis and Prediction*, Stanford, Calif.: Stanford University Press.

500. Sorum, M. (1971). "Estimating the conditional probability of misclassification," *Technometrics*, 13, p. 333.

501. Sorum, M. (1972). "Three probabilities of misclassification," *Technometrics*, 14, pp. 309–316.

502. Sorum, M. (1972). "Estimating the expected and the optimal probabilities of misclassification," *Technometrics*, 14(4), pp. 935–943.

503. Specht, D. F. (1967). "Vectorcardiographic diagnosis using the polynomial discriminant method of pattern recognition," *IEEE Trans. Biomed. Eng.*, BME-14, pp. 90–95.

504. Specht, D. F. (1967). "Generation of polynomial discriminant functions for pattern recognition," *IEEE Trans. Electron. Computers*, EC-16, pp. 308–319.

505. Specht, D. F. (1970). "Discrimination power of multiple ECG measurements," in *Clinical Electrocardiography and Computers*, New York: Academic Press, Inc., pp. 329–338.

506. Specht, D. F. (1971). "Series estimation of a probability density function," *Technometrics*, 13, p. 409.

507. Sproul, A., and Huang, N. (1966). "Diagnosis of heterozygosity for cystic fibrosis by discriminatory analysis of sweat chloride distribution," *J. Pediat.*, 69, p. 759.

508. Srivastava, J. N., and Zaatar, M. K. (1972). "On the maximum likelihood classification rule and its admissibility," *J. Mult. Anal.*, 2, pp. 115–126.

509. Srivastava, M. S. (1967). "Comparing distances between multivariate populations—the problem of minimum distances," *Ann. Math. Stat.*, 38, pp. 550–557.

510. Srivastava, M. S. (1967). "Classification into multivariate normal populations when the population means are linearly restricted," *Ann. Inst. Stat. Math.*, Tokyo, 19, pp. 473–478.

511. Stern, E., Crowley, L. G., Weiner, J. M., Hopkins, C. E., and Marmorsten, J. (1966). "Correlation of vaginal smear patterns with urinary hormone excretion," *Acta Cytol.*, 10, pp. 110–117.

512. Stevens, W. L. (1945). "Analyse discriminante," *Questoes Met. Inst. Anthropol. Coimbra*, 7, pp. 5–54.

513. Stilmant, M. M., Vamecq, G. M., Piessens, W. F., and Badjou, R. R. (1971). "Evaluation of extent of metastic liver disease proposed discriminant," *European J. Cancer*, 7, pp. 87–94.

514. Stocks, P. (1933). "A biometric investigation of twins, Part II," *Ann. Eugen.*, 5, pp. 1–55.

515. Stoller, D. S. (1954). "Univariate two-population distribution free discrimination," *J. Amer. Stat. Assoc.*, 49, pp. 770–777.

516. Stone, F. C. (1947). "Notes on two darters of the genus *Bolesoma*," *Copeia*, 2, pp. 92–96.

517. Sutcliffe, J. P. (1965). "A probability model for errors of classification. I. General considerations," *Psychometrika*, 30, pp. 73–96.

518. Sutcliffe, J. P. (1965). "A probability model for errors of classification. II. Particular case," *Psychometrika,* 30, pp. 129–155.
519. Sutton, R. N., and Hall, E. L. (1972). "Texture measures for automatic classification of pulmonary disease," *IEEE Trans. Computers,* C-21, pp. 667–676.
520. Tallis, G. M. (1970). "Some extensions of discriminant function analysis," *Metrika,* 15, pp. 86–91.
521. Takakura, S. (1962). "Some statistical methods of classification by the theory of quantification," *Ann. Inst. Stat. Math.,* Tokyo, 9, pp. 89–105.
522. Tarter, M., and Kronmal, R. A. (1970). "On multivariate density estimates based on orthogonal expansions," *Ann. Math. Stat.,* 41, pp. 718–722.
523. Tarter, M., and Raman, S. (1971). "A systematic approach to graphical methods in biometry," *Proc. 6th Berkeley Symp. Math. Stat. Probl.,* IV, pp. 199–222. Berkeley: University of California Press.
524. Tatsuoka, M. (1970). "Discriminant analysis. The study of group differences," Inst. Personality and Ability Testing, Champaign, Ill.
525. Tatsuoka, M., and Tiedeman, D. V. (1954). "Discriminant analysis," *Rev. Educ. Res.,* 24, pp. 402–420.
526. Teichroew, D. (1961). "Computation of an empirical sampling distribution for the *W*-classification statistic," in *Studies in Item Analysis and Prediction,"* H. Solomon, ed., Stanford, Calif.: Stanford University Press, pp. 252–275.
527. Tiedeman, D. V. (1951). "The utility of the discriminant function in psychological and guidance investigations," *Harvard Educ. Rev.,* 21, pp. 71–79.
528. Tomassone, R. (1963). "Applications des fonctions discriminantes a des problèmes biométriques," *Ann. Ecole Nat. Eaux et Forêts* 20, pp. 583–617.
529. Tou, J. T., ed. (1967). *Computer and Information Sciences,* New York: Academic Press, Inc.
530. Toussaint, G. T. (1972). "Feature evaluation criteria and contextual decoding algorithms in statistical pattern recognition," Ph.D. dissert., Dept. Elec. Eng., University of British Columbia.
531. Toussaint, G. T. (1972). "Polynomial representation of classifiers with independent discrete-valued features," *IEEE Trans. Computers,* C-21, pp. 205–208.
532. Toussaint, G. T., and Donaldson, R. W. (1970). "Algorithms for recognising contour traced handprinted characters," *IEEE Trans. Computers,* C-19, pp. 541–546.
533. Toussaint, G. T. (1974). "Bibliography on estimation of misclassification," to appear in *IEEE Trans. Inform. Theory.*
534. Travers, R. M. W. (1939). "The use of a discriminant function in the treatment of psychological group differences," *Psychometrika,* 4, pp. 25–32.
535. Truett, J., Cornfield, J., and Kannel, W. (1967). "A multivariate analysis of the risk of coronary heart disease in Framingham," *J. Chronic Diseases,* 20, pp. 511–524.
536. Tyler, F. T. (1952). "Some examples of multivariate analysis in educational and psychological research," *Psychometrika,* 17, pp. 289–296.
537. Uematu, T. (1959). "Note on the numerical computation in the discrimination problem," *Ann. Inst. Stat. Math.,* Tokyo, 10, pp. 131–135.
538. Uematu, T. (1964). "On a multidimensional linear discriminant function," *Ann. Inst. Stat. Math.,* Tokyo, 16, pp. 431–437.

539. Ullman, J. R. (1969). "Experiments with the n-tuple method of pattern recognition," *IEEE Trans. Computers*, C-18, pp. 1135–1137.
540. Urbakh, V. Y. (1971). "Linear discriminant analysis: loss of discriminating power when a variate is omitted," *Biometrics*, 27, p. 531.
541. Van Ryzin, J. (1966). "Bayes risk consistency of classification procedures using density estimation," *Sankhya*, A28, pp. 261–270.
542. van Woerken, A. J. (1960). "Program for a diagnostic model," *IRE Trans. Med. Electron.*, ME-7, p. 220.
543. van Woerken, A. J., et al. (1961). "Statistics for a diagnostic model," *Biometrika*, 17, pp. 299–318.
544. Vissac, B., and Wagner, R. (1963). "Analyse discriminante progressive, D^2 limitée à deux populations," *Station Centr. Gen. Animale.*
545. Von Mises, R. (1945). "On the classification of observation data into distinct groups," *Ann. Math. Stat.*, 16, pp. 68–73.
546. Wagner, R. (1965). "Progressive selection of variables using Mahalanobis' D^2 statistic. Application to determination of best discriminant function separating 2 seed populations of vine," *Annual Review of Plant Physiology*, 16, pp. 159–182.
547. Wagner, T. J. (1971). "Convergence of the nearest neighbour rule," *IEEE Trans. Inform. Theory*, IT-17, pp. 566–571.
548. Wagner, T. J. (1968). "The rate of convergence of an algorithm for recovering functions from noisy measurements taken at randomly selected points," *IEEE Trans. System Sci. Cybernetics*, SSC-4, pp. 151–154.
549. Wagner, T. J. (1973). "Deleted estimates of the Bayes risk," *Ann. Stat.*, 1, pp. 359–362.
550. Wald, A. (1944). "On a statistical problem arising in the classification of an individual into one of two groups," *Ann. Math. Stat.*, 15, pp. 145–162.
551. Wald, A. (1952). "Basic ideas of a general theory of statistical decision rules," in *Proc. Internatl. Congr. Math.*, American Mathematical Society, Providence, R.I.
552. Walker, S. B., and Duncan, D. B. (1967). "Estimation of the probability of an event as a function of several independent variables," *Biometrika*, 54, pp. 167–179.
553. Walsh, J. E. (1963). "Simultaneous confidence intervals for differences of classification probabilities," *Biom. Z.*, 5, pp. 231–234.
554. Wartmann, Rolf (1951). "Die statistische trennung sich in mehreren merkmalen uberlappender individuengruppen (diskriminanzanalyse)," *Z. Angew. Math. Mech.*, 31, pp. 256–257.
555. Watanabe, S. (1969). *Methodologies of Pattern Recognition*, New York: Academic Press, Inc.
556. Watanabe, S., ed. (1972). *Frontiers of Pattern Recognition*, New York: Academic Press, Inc.
557. Watson, H. E. (1956). "Agreement analysis: a note on Professor McQuitty's article," *Brit. J. Stat. Psychology*, 9, pp. 17–20.
558. Weber, A. A. (1951). "Efficacité de l'indice nasal comparée à l'analyse discriminante," *Acta Genet. Stat. Med.*, 2, pp. 351–363.
559. Weber, E. (1957). "Betrachtungen zur diskriminanzanalyse," *Z. Pflanzenzuchtung*, 38, pp. 1–36.

560. Wee, W. G. (1968). "Generalized inverse approach to adaptive multiclass pattern classification," *IEEE Trans. Computers*, C-17, p. 1157.

561. Wee, W. G. (1972). "On computational aspects of properties of a pattern classifier," *5th Hawaii Internatl. Conf. Systems Sci.*, pp. 598–600. Western Periodicals Company.

562. Wee, W. G., and Fu, K. S. (1968). "An adaptive procedure for multiclass pattern classification," *IEEE Trans. Computers*, C-17, pp. 178–182.

563. Wegman, E. J. (1972). "Nonparametric probability density estimation," *Technometrics*, 14, pp. 533–546.

564. Weiner, J., and Dunn, O. J. (1966). "Elimination of variates in linear discrimination problems," *Biometrics*, 22, p. 268.

565. Weiner, J. M., and Marmorsten, J. (1969). "Statistical techniques of difference," *Ann. N.Y. Acad. Sci.*, 161, pp. 641–668.

566. Weiner, J. M., Marmorsten, J., Stern, E., and Hopkins, C. (1966). "Urinary hormone metabolites in cancer and benign hyperplasia of prostate—a multivariate statistical analysis," *Ann. N.Y. Acad. Sci.*, 125, pp. 974–983.

567. Welch, B. L. (1939). "Note on discriminant functions," *Biometrika*, 31, pp. 218–220.

568. Welch, P., and Wimpress, R. S. (1961). "Two multivariate statistical computer programs and their application to the vowel recognition problem," *J. Acoust. Soc. Amer.*, 33, pp. 426–434.

569. Wesler, O. (1959). "A classification problem involving multinomials," *Ann. Math. Stat.*, 30, pp. 128–133.

570. Whitney, A. W. (1971). "A direct method of nonparametric measurement selection," *IEEE Trans. Computers*, C-20, pp. 1100–1103.

571. William, J. H., Jr. (1963). "A discriminant method for automatically classifying documents," *Proc. American Federation of Information Processing Societies, Fall Joint Computer Conference*, New York: Spartan Books.

572. Williams, E. J. (1955). "Significance tests for discriminant functions and linear functional relationships," *Biometrika*, 42, pp. 360–381.

573. Williams, E. J. (1961). "Tests for discriminant functions," *J. Austral. Math. Soc.*, 2, pp. 243–252.

574. Wingnall, T. K. (1969). "Generalized Bayesian classification functions: K classes," *Econ. Geol.*, 64, pp. 571–574.

575. Wolverton, C. T., and Wagner, T. J. (1969). "Asymptotically optimal discriminant functions for pattern classification," *IEEE Trans. Inform. Theory*, IT-15, pp. 258–265.

576. Wolverton, C. T. (1973). "Strong consistency of an estimate of the asymptotic error probability of the nearest neighbour rule," *IEEE Trans. Inform. Theory*, IT-19, pp. 119–120.

577. Woo, T. L., and Morant, G. M. (1932). "A preliminary classification of Asiatic races based on cranial measurements," *Biometrika*, 24, pp. 108–134.

578. Yardi, M. R. (1946). "A statistical approach to the problem of chronology of Shakespeare's plays," *Sankhya*, 7, pp. 263–268.

579. Yau, S. S., and Schempert, J. M. (1968). "Design of pattern classifiers with the updating property using stochastic approximation techniques," *IEEE Trans. Computers*, C-17, pp. 861–872.

Index